Child-Robot Interaction

Methods, Epistemology, Applications

Child-Robot Interaction

Methods, Epistemology, Applications

Edited by
Silvia Larghi
Edoardo Datteri

ISBN 978-1-84890-466-8

College Publications
Scientific Director: Dov Gabbay
Managing Director: Jane Spurr

http://www.collegepublications.co.uk

Cover produced by Laraine Welch

This book has been published with the
contribution of the University of
Milano-Bicocca – "Riccardo Massa"
Department of Human Sciences for
Education

Contents

Preface

Robotics and artificial intelligence technologies are advancing at a rapid pace, enabling forms of direct, physical and cognitive interaction between robots and humans in a wide range of fields, from healthcare to education, from entertainment to personal assistance. Increasingly sophisticated, even humanoid, robots can now safely interact with vulnerable people - especially children - and serve as facilitators of teaching, learning, rehabilitation and therapeutic processes. These developments need to be critically monitored and reviewed. Robots can make a significant contribution to improving the quality of certain processes, but it is essential to identify the areas in which they can be more useful and to establish methodological and ethical guidelines for their use. With regard to education and training, it is important to reflect on the specific role they could play - peers, tutors, programmable tools? - and what benefits they can really bring to the field. What disciplinary and cross-curricular skills and competences can be developed using robots? What is it about robots that makes them more 'useful' in some interesting sense than other kinds of educational tools? How can learning objectives achieved with robots be extended and generalised to other domains? How are robots perceived and understood by teachers, educators and students, also depending on their cultural background?

These questions cannot be addressed solely from the perspective of robotics research and development. They require collaboration between roboticists, psychologists, philosophers, anthropologists, educators, teachers, pedagogists and health professionals. Robotics is inherently multidisciplinary, and it is becoming increasingly clear that the fields that need to work together to deeply understand the reasons, the meaning, the significance, the implications of using robots in contexts frequented by children are not limited to science and technology in the traditional sense. The time has come to seriously consider the perspective of the so-called humanities, which could make a decisive contribution to understanding the dynamics of child-robot interaction, to identifying the real benefits of these technologies in terms of cultural and educational progress, and to opening up innovative and human-centred avenues of research.

In order to create such a *milieu*, the RobotiCSS Lab (Laboratory of Robotics for the Cognitive and Social Sciences) of the University of Milano-Bicocca organises every two years the International Conference on Child-Robot Interaction. It is a generalist conference covering practically all aspects of child-robot interaction, with the aim of facilitating the cross-fertilisation of different disciplines, languages and theoretical backgrounds. The organisers are proud to reflect this multidisciplinary character in the structure of the RobotiCSS Lab, which is essentially a robotics laboratory that has grown up in a humanistic university department, composed of researchers with different backgrounds - philosophical, psychological, pedagogical, anthropological - and dedicated to the study of the human side of human-robot interaction.

This book contains the full text of some of the papers presented at the 2023 International Conference on Child-Robot Interaction (CRI23), held in Milano-Bicocca from 23 to 25 June. The Conference was organised in collaboration with INDIRE (Istituto Nazionale Documentazione Innovazione Ricerca Educativa) and the Università Politecnica delle Marche. It is divided into two parts. The first, from chapters 1 to 8, contains empirically based but theoretically oriented contributions on the role of robots in education and healthcare and on users' perceptions of them. The second, from chapters 9 to 14, includes papers presenting empirical research on the role of educational robots in supporting science and language learning, and in promoting nutrition education. All the papers were subjected to double blind peer-review. We gratefully thank the reviewers: Alex Barco Martelo, Gilda Bozzi, Cristina Caldiroli, Antonio Chella, Augusto Chioccariello, Valeria Cotza, Luisa Damiano, Silvia Di Battista, Margherita Di Stasio, Marta Díaz Boladeras, Amy Eguchi, Lorella Giannandrea, Renato Grimaldi, Hagen Lehmann, Gabriel Lemkow Tovias, Petar Vasilev Lefterov, Jose Mangione, Beatrice Miotti, Anastasia Misirli, Lucio Negrini, Giovanni Nulli, Fiorella Operto, Monica Pivetti, Laura Screpanti, Luca Zanetti, Luisa Zecca. We also express our gratitude to the members of the CRI23 organizing committee: Gilda Bozzi, Margherita Di Stasio, Laura Messini, Beatrice Miotti, David Scaradozzi, Laura Screpanti, and Luisa Zecca.

We gratefully acknowledge financial support by the Department of Human Sciences for Education of the University of Milano-Bicocca, and the Rectorate of the same University.

In anticipation of the next CRI conference in 2025, we hope that this book will make some progress towards a critical and constructive reflection on if and how robots can improve the quality of children's lives and promote their personal growth.

Silvia Larghi, Edoardo Datteri

Part I

Reflections on the roles of robots in education

Embodied learning: biological-artificial hybridization

Rosa **Gallelli***[1]*, Loredana **Perla***[2]*, Angela **Balzotti***[3]*, Stefania **Massaro***[4]*, Viviana **Vinci***[5]* and Pasquale **Renna***[6]*

[1] University of Bari, via Crisanzio, Bari, Italy, 0000-0001-7765-5628, rosa.gallelli@uniba.it
[2] University of Bari, via Crisanzio, Bari, Italy, 0000-0003-1520-0884, loredana.perla@uniba.it
[3] University of Bari, via Crisanzio, Bari, Italy, 0009-0006-9295-3125, angelabalzotti@tiscali.it
[4] University of Bari, via Crisanzio, Bari, Italy, 0000-0003-4695-1007, stefania.massaro@uniba.it
[5] University of Foggia, via Arpi, Foggia, Italy, 0000-0002-4091-0098, viviana.vinci@unifg.it
[6] University eCampus, via Isimbardi, Novedrate (CO), Italy, 0000-0001-8272-6814, pasquale.renna@uniecampus.it

Abstract

Starting point of the proposed reflection is the hypothesis that processes of subjectivation are marked by an irreducible tension towards what Rosi Braidotti (2014) has defined as a "posthuman becoming" understood as a process of openness and hybridization of the subject towards the otherness of nonhuman, organic and machinic entities to which the subject incessantly connects. (Maturana & Varela, 1992; Marchesini, 2002, 2009, 2014; Pinto Minerva&Gallelli, 2004). In particular, studies on complexity and autopoiesis have indicated that the peculiar characteristics of living systems are: knowledge as an evolutionary capacity linked to the exchange of information and autopoiesis as the ability of a complex system to use knowledge in order to: a) maintain the organization scheme of its components; b) keep its structure open to the exchange of information and energy with the environment. So, it is possible to say that life itself is a process of knowledge that allows an autopoietic system to adapt and survive. The novelty introduced by the hybridization between biology and technology is, on the one hand, the incorporation of technologies into biological systems and, on the other hand, the grafting of biological characteristics into machinic systems: the ability of machines to learn, to repair themselves, to self-regulate, even to evolve, autonomously (Kelly, 1996). Social robots represent one of the most advanced frontiers of the hybridization processes just described. In fact, research in this field is focused on the possibility of equipping such anthropomorphic robots with the ability to acquire knowledge and information from the environment, to connect constructively previous knowledge and acquired knowledge, and to activate relationships with humans and other robots. The transformative power of the convergence between biological and artificial can be fully understood in the light of the various embodiment theories (Merleau-Ponty, 1965; Varela, 1994; Barone, 2014; Gallelli, 2017) which – in multiple and different scientific fields – have corroborated the pioneering theses advanced in the 1960s by McLuhan on the aspects of interdependence between cognitive functions and technological artifacts and have explored the specific role played by the body and by action in this interdependence. On the one hand, the functions of thought appear inextricably linked to the possession of a body. On the other hand, each technology determines diversified kinds of thought to the extent that it opens up specific spaces of perceptual-cognitive action. This implies a series of crucial questions that raise questions on an ethical and pedagogical level: 1. What evolution will the cognitive apparatus of social robots undergo as they will be able to count on "bodies" increasingly capable of interacting with the world, to learn and build relationships? 2. What evolution will human subjects undergo as the world in which they move and evolve will be populated by machinic entities to build knowledge and relationships with? 3. Will the relationship between humans and robots be able to assume autopoietic characteristics within the virtual environments of the technological systems of today, where both humans and machines modify their own structures and capacities and co-implicate themselves in an itinerary of changing?

Keywords
Embodied Learning, Social Robot, Posthuman

1. Pedagogical studies on the post-human and on the human-robot relationship

The pedagogical and didactic topic connected to the multiple implications of "embodied learning", with the related hybridizations between biological and artificial, up to the most recent interactions between human and robot, has been problematized by authoritative contemporary academic pedagogists.

Alessandro Peter Ferrante (2014), in this regard, highlights how the power of contemporary technology, by retroacting on the human beings who experience it, produces reciprocal structural modifications.

Technique, therefore, especially today, can no longer be conceived simply as a docile tool in the hands of humanity, without being altered in any way. In fact, it retroacts directly and indirectly on man himself. Moreover, it delimits, shapes and models his morphological structures, his cognitive and emotional faculties, his social and environmental relationships, the forms of his acting and experiencing: man builds his tools and tools help to build man (Longo, 2006, p.

86). Man is not only the architect and producer of the various techniques, but he is also their product, so much so that a philosopher, Peter Sloterdijk, emphasizing this aspect, has focused on the role exercised in the construction of the human by the various techniques, explicitly defined as *anthropotechnical*, that is to say techniques for the formation of the body and identity of man (Sloterdijk, 2001). Contemporary techniques generate a significant gap than even before. They are not limited only on modifying the structures of experience and the conditions of existence, but a l s o o n creating the conditions for the abandonment, the criticism, the remodulation of the most common representations that human beings in the Western tradition have elaborated on themselves. (Ferrante 2014, p. 18) [1]

A peculiar kind of technique hybridized with the human is represented by the robot, as an artificial humanoid capable of imitating and enhancing the cognitive and now also affective faculties of the human.

Robot therefore start from an artificial body, equipped with sensors and actuators. It is equipped with artificial intelligence and, in a more or less distant future, with artificial emotions and, lastly, perhaps, with an artificial conscience. Furthermore, the robot is already characterized by a relative autonomy and by a relative learning capacity, which make it a plausible candidate for an evolution [...] of both a humanoid and an alternative type to the human. The imitative evolution of the human could lead to machines indistinguishable from us for functions (intellectual, active, perceptive, emotional...) even if distinguishable for materials and partly for structure and appearance. [...] Through the pouring of our minds into the artificial intelligences of robots, according to the perspective of the "children of the mind" outlined by Minsky, robots could collect our inheritance and fully re-enter the post-human vision. (Longo, 2014, p. 96)

The robot today represents not only a concrete possibility of merging studies on artificial intelligence and those on prosthetic machines, but also an applicative possibility of experimenting with the limits and the applicative potential of "intelligent" programs, where intelligence is configured, in the biological field, as a property emerging from the multiple spatio-temporal interactions between living and non-living people and, in the technological field, as a property emerging from the "situated" interactions of artificial intelligences applied to robotics.

Franca Pinto Minerva and Rosa Gallelli (2004) have elaborated significant reflections on some peculiarities specifically related to the educational potential of "intelligent programs". In fact, in the essay "Pedagogy and post-human. Hybridization of identities and frontiers of the possible" they underline how, since the early 2000s (Pinto Minerva&Gallelli, 2004), pedagogical and didactic knowledge in Italy has developed a specific and constant interest in the topic of the interaction between human and machine, and increasingly between biological knowledge (and life biology connected

---

[1] All the quotes in the text are translated by Italian into English by the authors of this essay.

to it) and knowledge about artificial life, starting from the educational field linked to the evolutionary developments of the co-implications of the living-system and the machine-system.

In the aforementioned essay, the authors recall how knowledge about biological life

> has traditionally adopted an analytical type of investigation methodology: starting from the organism in its biochemical unity and complexity and breaking it down (top-down, from top to bottom) into its increasingly smaller components (organs, tissues, cells, molecules), it has put together the general framework of the "mechanics" of life on Earth. [Knowledge related to] artificial life, by contrast, takes a synthetic approach to its research object. [...] Far from limiting herself to the study of the *mechanics* of organisms, she turns instead to studying the *dynamics* of these: their principles of self-organization [...]. This bearing in mind the fact that knowledge related to artificial life is characterized by a "bottom-up approach whereby global behavior emerges from the set of local activities without being represented by rules placed in a particular point of the system. (pp. 64-65)

Moreover, upstream of these topics there is the question of "embodied" intelligence between "natural" and "artificial", particularly related to the interdependence between cognitive functions and technological artefacts, and the specific role played by the "body" in this interdependence.

In fact, on the one hand the functions of thought appear inextricably linked to the possession of a body. On the other hand, each technology determines diversified kinds of thought to the extent that it opens up specific perceptual-cognitive application spaces.

Taking into consideration the first aspect of the topic, i.e. the link between mental functionality and the body size of the subject, this means:

1. That rational and symbolic activity is not the only form of cognitive activity;
2. that symbolic activity exists in its current form only as it has evolved in the close (and inseparable!) interconnection with the material substrate of the body.

About the first point, in addition to the interpretation of intelligence exclusively as a symbolic and reconstructive capacity – an interpretation that has its roots in the Western philosophical tradition – another vision has established itself in the field of cognitive psychology. Cognitive activity, in fact, appears to be configured in two very different types of functioning.

On the one hand, there is the manipulation of symbols – for a long time explicit, conscious, intentional – which finds in language both a very powerful amplifier and an extraordinary product. Dealing with symbols, the mind has to perform two important operations: firstly the decoding the symbols, i.e. extracting their meaning, and secondly the reconstruction of the more general situation to which the succession of symbols refers. On the other hand, there is the way of functioning – largely implicit, unaware and automatic – of perceptual – motor learning.

The latter appears to be closely related to the action of the subject within an environment: objects are known not simply by perceiving them, but above all by correlating the object's reaction to one's own motor action. There is knowledge, in other words, when the right correlation is established between the object's response (which is learned through perception) and the action that generated it. In the case of symbolic-reconstructive learning, the cognitive process makes use of the "mediation" of symbols and does not imply any exchange with the outside except, precisely, that with the symbols themselves: therefore, it exclusively involves visual perception and requires the slow processing of the information to the extent that they are continually subjected to the double interpretative/reconstructive step.

Instead, in the case of perceptual-motor learning, the cognitive process implies the experience of the reality to be known, i.e. an immediate and continuous exchange of perceptive inputs and motor outputs with the outside world: therefore, it necessarily involves the total sense- perceptive and kinesthetic of the subject and puts into play a rapid processing of information to the extent that they derive simultaneously from different channels, in the context of repeated cycles of perception- action.

> The transformative possibilities opened up by the technological evolution in progress - with the new forms of integration they solicit between cognitive dimensions and bodily dimensions, between logical-symbolic dimensions and perceptual-motor dimensions - appear confirmed and re- launched by the most recent research in the field of robotics. These researches, in fact, are developed by giving importance precisely to the perceptive-motor capacities to the extent that, at the basis of intelligent behavior, the complex system of perceptive input- motor output is recognized which is established when an "agent" moves - with its own "body" - within an environment, experiencing it on the basis of interests and motivations, intentions and purposes. [...] Already in the mid-eighties, in fact, robotics appeared to be able to take charge of the problems that Artificial Intelligence had run into, to face them with a very different approach. In fact, it shifts the attention from a strictly computational reading of replication to the design and construction of real systems, i.e. systems capable of moving in unstructured environments – "dirty" - just as natural organisms are capable of doing. [...] Attention is [...] focused on the mechanisms of learning and recognition, and on how even intelligent behavior "emerges" from the complex and dynamic relationship that the organism weaves with the surrounding environment through its own body. (Pinto Minerva&Gallelli 2004, pp. 86-87, p. 89)

2. *Status quaestionis*. Embodied mind and artificial bodies. A theoretical analysis on the possibilities of a meeting between Intelligence and Robot

Today, pedagogical studies on the interaction between humans and robots move from a theoretical background of Darwinian matrix.

Studies of a Darwinian nature support the formative possibility of an orientation of the evolution of living systems. On these studies are grafted those of a neuroscientific nature, which argue the close relationship, as has been said, between mind and body in a body-mental unity.

The development of neuroimaging techniques is making a significant contribution to the possibility of understanding the evolutionary peculiarities of the human mind. The neurobiology of emotions argued by the neuroscientist Antonio Damasio (1995) offers interesting clues about the conditions of the emergence of human specificity starting from a fundamental sharing with other animals of adaptive behaviors rooted in the same deep bodily needs. Indeed, for the scholar, there is a close continuity between the emotional and bodily skills of organisms equipped with automatic, unlearned action programs aimed at protecting their own lives and regulated by homeostasis and the cognitive faculties of our species, in which emotions mobilize resources and strategies directing them towards the management (which may be conscious) of behaviors aimed at the protection of life.

Precisely in relation to this difference, Damasio highlights the inextricable relationship between the development of the mind in a given subject and the possession, by this subject, of the ability to perceive his own body moving in the environment:

In fact, not being able to move means not having a behavior. And without behavior it is not possible to talk about the mind. In a sense, the minimum requirement for an organism to develop consciousness is that it has a brain, that it can move, that it has perception of its body and the external environment and that it builds a policy of reaction to stimuli - which can be called behavior - initially based on a series of reflections (Damasio in Pievani, 2008, p. 50). Along the same lines, Harvard psychologist Steven Pinkler argues that humans evolved their intelligence to fill an evolutionary niche in the ecosystem, manipulating the environment through causal reasoning and cooperation. Alongside this, for the author "there is a process of metaphorical abstraction, still visible today in our language, which allows cognitive mechanisms suitable for physical reasoning to adapt to abstract elements" (Pinlker in Pievani, 2008, pp. 61- 62).

The concept of "cognitive niche" implies the idea that, if it is true that all organisms evolve one at the expense of the other, through processes of selection and mutation, it is also true that the human species has participated in these processes using the reasoning of the cause-and-effect type from whose "internalization" specific models of functioning of reality derive. In other words, by entering a cognitive niche, we could manipulate the world to our advantage. Cognition in all its forms, from the most automatic to the most complex, consists in progressively learning to connect the different perceptions to their favorable or harmful effects on the survival conditions of the organism, producing adequate synaptic modifications. During the development of the child, therefore, the experiences that take place in a given environment act as "specialists" in the sense of pruning, within the redundant number of nerve cells and synaptic possibilities available at birth, certain particular neuronic groups, organizing them in maps of groups, each one precisely specialized in certain tasks. In the context of these theories, therefore, the construction of the brain of each individual subject - and of the mind that originates from it - is the absolutely singular fruit of a selective process originating from the intersection between a genetically determined structure and a 'cultural imprint' impressed by experiences of the subject on the cerebral system.

From the same perspective, i.e. that of linking the mind to a certain cerebral morphology, in turn produced by selection, Edelman moved to develop a biological theory of consciousness that could describe the origin of its properties both in terms of and permanent neuronal functions and in terms of evolutionary events. The resulting image of the brain is clearly in contrast with that of a computational machine which processes the incoming information according to a set of instructions or a pre-set code and which finally produces an answer in an unambiguous way. The brain, on the other hand, is a complex and dynamic system that emerges from a selective interaction with the world. (Pievani, 2008, pp. 151-152, 154) However, Edelman puts forward a bold hypothesis for the 1990s, and yet one that is current today. For the scholar, in fact, the advance of knowledge on the human brain would soon lead to the creation of artificial beings.

> As soon as we know with greater certainty the facts and principles relating to the structure and function of the brain, [...] we will construct an artifact endowed with consciousness and, if we are able to give it the basis of language, we may ask whether it categorizes [...], as we do. (Edelman, 1999, p. 177)

As previously argued, neuroscientific studies have posed the problem of the emergence of intelligence from the surrounding environment. On the basis of Edelman's work, as well as the theoretical proposals of George Lakoff, Mark Johnson, Eleanor Rosh and Francisco Varela, Damasio in Descartes' error proposes an interesting interpretation of the bodily bases of cognition and, more precisely, he finds the biological basis of learning. Not only do learning processes fall within the biological process through which our species evolved, but also learning processes, in their original evolutionary value, appear rooted in the entire human body. Damasio (1995, p. 318) conceives the mind as 'incorporated'. The mind, in this perspective, coincides with the set of representations of which it is possible to acquire awareness in the form of images. The primitive representations of the body in activity would provide us with a temporal and spatial frame, a metric to which to relate the other representations. The most authoritative twentieth-century studies of a phenomenological nature underline the importance of the perception of this environment as an indivisible unicum. The representation of what we now construct as a three-dimensional space would be generated in the brain, based on the autonomy and movement patterns of the body in the environment (Damasio, 1995, p. 319). Ultimately, what we know of external reality, and even the awareness we have of it, is achieved by the body in activity, through the representations of the perturbations caused to the body itself by the environment with which it interacts. Indeed, Damasio warns that the representation of the body is continuously ongoing even when the subject's attention is consciously directed to representations of external events (through sight, hearing, touch) or to internally generated images. The understanding of the mind and of the interconnection between cognitive dimensions and body dimensions in the definition of the special human individuality, as filtered by the cognitive sciences, appears exposed to the difficulties of a "third person description" which obviously tends to "objectify" the fulcrum of the investigation cognitive.

These are the difficulties that the phenomenological reading of Maurice Merleau-Ponty (1979) reveals when he questions the discrepancy between Leib and Korper: living body and anatomical body. In deepening the phenomenology of Husserl (1931), Merleau-Ponty manages to develop the conceptual core of his own philosophy of corporeity, aimed at unhinging the mind-body dichotomy and advancing the idea of their synergistic, continuous interdefinition and interdependence. For him, placing ourselves in the presence of the "world of life", i.e. of the "world in which we live prior to reflection", allows Husserlian philosophy to lead us to the awareness of the "wild residue" of every reflection: the experience that the subject makes of the world "before" and beyond any reflection, when it simply "exists". In the naturalness of this existence, the subject experiences an original belonging-participation in the world independently of the judgmental and propositional acts that reflection produces at a later time. The "natural" and "wild" experience evoked by phenomenology is the experience of the subject as - above all - Leib (living body). Merleau-Ponty here refers to the notion of Leib expressed by Husserl: proper body, point of reference with respect to which the pre-reflexive coordinates of each movement of the subject (task, choice, behavior, etc.) are structured (Gallelli 2014).

> The deepening of the notion of Leib expressed by Husserl leads Merleau-Ponty to delineate the living body as an integrated reality of rationality and materiality, a structurally "ambiguous" reality: to the extent that it is not pure materiality that is visible, objectifiable and separable from the invisible and rational

consciousness. The living body is already, prior to any predicative operation, a body located in a space/time environment (a physical, historical and social environment), with respect to which it spontaneously develops non-decisive and unexpressed "intentions". The latter are inherent in the ability of the Leib to organize and structure the stimuli of the environment starting from his own reality of absolute "here". Without a doubt, the author is influenced by the conceptualizations advanced by Gestalt psychology on perception. The latter, in fact, in that historical period had attributed to perception the characteristics of globality and significance, which decisively put into check the typical dichotomy of nineteenth-century analytical psychology between the merely receptive role of the biological organism (reception of discrete and localizable sensitive data) and, on the other hand, the cognitive role of intelligence which establishes links between data, and organizes them into meaningful configurations. From this perspective, perceptive activity is, for Merleau-Ponty, always already an activity that spontaneously involves individual subjectivity in its entirety: my perception is therefore not a sum of visual, tactile, auditory data; I perceive in an undivided way with my total being, I grasp a single structure of the thing, a single way of existing that speaks to all my senses at the same time (ibid., p. 177). (Gallelli 2014, pp. 155-156)

Studies of a phenomenological nature were then corroborated by studies of a biological nature which underline how the categorizing attitude of the Western intellect, in this sense, does not account for the autopoietic, self-organizing function of living systems, which are capable to maintain a situation of equilibrium based on reciprocity processes in the context of systemic interaction. Every act of perception is crossed by intentions, evaluations and expectations on the part of the subject who, far from dividing up the perceptive field in an aseptic way, using the different sensory channels in a sectorial way, grasps the objects in their richness of meaning and included in a context to which a personal emotional tone is attributed. Taking up the analysis developed by Merleau-Ponty, the Chilean neurobiologist Francisco Varela (1992) proposes the notion of knowledge as an enation. The term refers to the ability, intrinsic to all organisms, to generate the meanings of reality from within and to do so on the basis of the constraints and possibilities of its physiology: of its body, starting with its sensory-motor structure. The construction of knowledge systems, in this sense, is conditioned by the way in which each organism organizes its action in the world (and on the world) as well as by the course in which these actions unfold in the historical time of that organism's ontogeny (and in the historical time of the phylogeny of the species to which it belongs). Cognition, therefore, far from being defined as the result of a "mind" that "captures and represents" the aspects of a reality that appears in front of it as a pre-existing object in its own right, derives from the encounter between an organism that moves and acts in the environment - with the peculiar ways that its "body" makes available to it - and the peculiarities of the environment itself: Perception does not consist in the recovery of a pre- existing world, but rather in the perceptive guide of the action in a world that is inseparable from our sensorimotor skills. Cognitive structures emerge from recurring patterns of perceptually guided actions. The cognition consists not of representations but of embodied action. So, the incorporated structures (sensory-motor) constitute the substance of experience and the experiential structures "motivate" conceptual understanding and rational thought (Varela, 1994, pp. 157-158).

The conceptual background of this conception in which a fundamental body-mind continuity is found is constituted by the conception of knowledge already proposed by Humberto Maturana (1985) and then further explored by Maturana and Varela. Cognition, according to Maturana, is inherent in every living system, understood as an autopoietic system. The general theory of systems, right from the first formulations given by the biologist Ludwig von Bertalanffy (1968), had already shifted cognitive attention from objects, simple and elementary units, to "complex units", made up of parts related to each other according to particular organization schemes. In short, each system presents a configuration of relationships between the parts, precisely regulated, which guarantees its identity beyond the transformations that the system also undergoes in the exchanges it maintains with the environment. In this framework, the concept of emergency stands out.

In fact, with respect to the properties and qualities of the individual elements of which it is composed, the system has specific properties that do not correspond to the sum of the properties of the elements themselves: the properties of the system "emerge" from its organizational structure which, binding the parts to rules and determinisms, involves new constraints and possibilities, unknown to the parts and proper to the "whole" to which they belong. Edgar Morin effectively writes, deepening this conception in his theory of complexity, that "emergence constitutes a logical leap, and opens in our intellect the passage through which irreducibility penetrates" (Morin, 2001, p. 123): the concept of emergence allows us to understand the impossibility of "reducing" certain aspects of reality to the elements of which that reality is composed, each aspect being the "emerging quality" from the overall organization of the system to which those elements belong (Annacontini, 2008).

In this perspective, for Maturana living systems present an organization scheme characterized by a reticular configuration of relations between its components of an autopoietic type to the extent that this internal organization is a network that continuously produces itself. The order and behavior of this organization scheme, in other words, do not depend on environmental influences and derive from autonomous processes of self-organization. On the other hand, living systems present an openness to the exchange of energy, matter and information with the environment which continuously modifies their physical structure.

In the course of life, therefore, every living system, in a complex process of integration between "closure" and "openness" to the environment, interacts with the other systems of its world, encountering "perturbations": solicitations that push it to trigger structural changes compatible with one's own autopoietic possibilities. Living systems and the environment thus appear to be interested in incessant co-evolutionary processes of reciprocal determination. In the model formulated by Maturana and Varela, beyond what we commonly consider "thought" - language, conceptualization, knowledge - cognition "emerges" from the material complexity of which every living system is kneaded and is configured as a process of knowledge that includes all the processes of transmission, exchange and management of information through which living systems are able to restructure the material conditions that guarantee them the stability of their own autopoietic organization.

For an autopoietic system, knowledge is a fundamental survival strategy and is, therefore, found in all forms of life, regardless of the presence of a real brain system. In this way, a new synthesis between "mind" and "matter", between "mind" and "body" is configured. All living organisms reveal how their cognitive processes - from elementary forms to human intelligence - are processes of self-organization, which are determined within the complex game of interconnections between different levels of organic life, inseparable from bodily experience: from the corporeity of the simplest biological phenomena up to the kinesthetic and sensorimotor corporeity of complex organisms. The main topic of this emerging perspective is the belief that the essence of knowledge is first and foremost concrete, embodied, incorporated, lived. This unique, concrete knowledge, its historicity and its context, is not the 'noise' that prevents the general model from being understood in its true substance, an abstraction, nor does it constitute a stage towards something else: it is the way where we operate and the place where we are (Varela, 1994, p. 144). The neuroscientist Domenico Parisi writes about it: "Behavior and mental life cannot be reduced [...] to nerve cells, body components and genes, since behavior and mental life are emergent properties of complex systems of which nerve cells, individual parts of the body and genes are the components. As in the case of all complex systems, it is not possible to predict or reduce the characteristics of behavior and mental life by knowing these components and their interactions (Parisi, 1999, p. 76). (Gallelli in Gallelli&Annacontini, 2014, pp. 160-166)

In the light of the above, we could ask whether the aforementioned theories of embodiment are able to represent a grid of intelligibility useful for interpreting the multiple implications of the transformative power of the convergence between "biological" and "artificial", in particular in the robot as an artificial subject able to self-regulate, self-repair, learn, experience emotions.

3. Robots and training planning

How do new robots learn? How can robots interact with humans? And how can robots understand and express human emotions? Trying, together with the girls and boys, to answer these questions allows us to face in a completely new way the sphere of intelligence, of the construction of knowledge, of motivations and decisions, as well as of affectivity and emotions, fundamental for human social interactions. Advances in neuroscience and cognitive science provide us with a new picture of how the brain works and what intelligence is: a picture in which, alongside cognitive processes, the fundamental role of emotions emerges more and more. The ability to recognize and manage one's own emotions and those of others form a properly emotional intelligence. In this type of intelligence includes emotional learning related to social relationships, which involves fundamental interpersonal skills, including cooperation, collaboration and communication. This type of skill, termed Social and Emotional Learning, plays a key role in academic achievement, the development of healthy social relationships, and the prevention of self-harming or antisocial behavior.

In fact, in a society characterized by growing complexity, affected by rapid and unpredictable changes in culture, science and technology, it is extremely necessary that young people possess not only theoretical knowledge and technical skills, but above all attitudes of openness towards innovations, availability for continuous learning, taking on independent initiatives, responsibility, flexibility and collaboration. In the face of this scenario, the setting up of an educational robotics laboratory combines the charm and topicality of its interdisciplinary contents centered on robots, which feed curiosity and passion for science, with the use of new learning methods centered on the philosophy of teaching skills: learning "in the field", built in a participatory, inclusive and cooperative way, attentive to the processes and not only to the contents, strongly aimed at balancing the cognitive, emotional-affective and relational aspects. Playing, designing and learning to program a robot, reflecting on the mechanisms that regulate intelligence and human social interactions, means developing skills and abilities that are useful not only from a technological/scientific point of view, but also from that of problem solving, emotional intelligence, creativity and teamwork.

The class operates as a real "community of scientific practices", with a strongly inclusive dimension, also useful for the purposes of integrating disabilities. The robot generates amazement and interest; solicits an emotional transference such as "being in need of care"; stimulates and maintains attention. In the MIUR programming document intitled: "For the school - skills and environments for learning", educational robotics gets full recognition: "the application of this methodology combines the application of computational thinking with a clear multidisciplinary approach that includes physics, mathematics, computer science, industrial design, as well as social sciences. Furthermore, due to the many fields of application, teamwork, creativity and entrepreneurial skills are required for the innovative design, programming and exploitation of robots and robotic services. Male and female students are attracted by these autonomous machines, which can become learning mediators and be a means of acquiring skills and a tool for sharing ideas". In addition to presenting itself as an innovative teaching methodology, educational robotics nonetheless fuels curiosity and passion for science, contributing to the strengthening of teaching and learning processes of scientific and humanistic subjects.

The STEM career guidance function is carried out with specific attention to the issue of equal opportunities, fully aware of the importance of involving and motivating female students as well as male students, providing them with tools and opportunities to deal with these subjects and eventually choose the area of these technical-scientific careers, in which Italy is called to fill a historic misalignment with respect to other countries. A last consideration, but no less important, is linked to the increasingly pervasive and dominant role of technology in society and the economy. As the "Decalogue of the rights of digital natives" reminds us, our pupils, our children, have the right to use technology without being used by it. From this point of view, setting up a robotics laboratory plays a decisive role in preparing, stimulating and accompanying students, often passive and unaware consumers, towards an understanding and use of digital technologies that manages to go under the surface, thanks to new interpretations. Having in mind the objective of "an ethical and responsible technological practice, far from inappropriate reductionisms or specialisms and attentive to the human condition in its entirety and complexity", educational robotics thus becomes, ultimately, also a tool for a practice of citizenship more critical, aware and creative with respect to the open problems to which the future is called to give an answer. (D.M. 254 del 16 november 2012; G.U. n.30 del 5 february 2013).

Below is an example of a robotics laboratory model based on the indications above.

Purposes of the laboratory. Introduction to social robotics Conception and design of the interactions of a social robot:

- To introduce knowledge of the fundamental principles and concepts of robotics
- Strengthen logical and problem-solving processes
- Introducing the knowledge of coding (block programming)
- Reflect together on ethical issues and the role of technology in society (digital rights)
- Improve understanding, analysis and problem-solving skills and critical evaluation of situations

- Motivating and teaching students to learn to learn (metacognition)
- Motivate students to investigate the relationship between cognition and emotion
- To introduce the knowledge of the fundamental principles and concepts of social robotics
- Formulate hypotheses on the nature of emotions (what are they? What are they for?) and on the sensory, perceptive and behavioral inputs that are able to arouse/inhibit them (what does the color white transmit to me?)
- Strengthen the ability to recognize and manage one's own emotions and those of others (emotional intelligence)
- Promote awareness of the role of emotions in decisions
- Activate an experiential dimension of learning
- Make operational the basic knowledge acquired on different types of emotions, on human social signals, and on non-verbal language.
- Create a "research community", a learning group capable of building a common research path through dialogue and argumentation.
- Orient the socio-emotional development and collaboration towards shared objectives in a positive sense
- Improve social-relational skills and assertiveness through presentations "in public".

Objective of the laboratory. In order to acquire the fundamentals of programming social robots, in order to conceive and implement, through group work, the setting and programming of a robot's journey to discover the six fundamental emotions.

Summary of contents.

- Robots and cognitive aspects of intelligence. First of all, the project aims to provide the broadest overview of the different types of robots (humanoid, mobile, home automation, service, companions) which in the coming years will forcefully enter everyday life, with an introduction to the basic components (sensors, actuators, etc) and with the simplified description of the mechanisms that regulate the "artificial intelligence" of robots (cognition, learning, motivation, emotion).
- Introduction to coding.
- Social robots and emotional aspects of intelligence. The project also aims to particularly intrigue children on the issue of social robots

(robots capable of interacting socially on the basis of emotional capacities). To this end, students will participate in experiential activities that allow them to learn firsthand the importance and social utility of emotions and to identify the specific mechanisms by which they can operate in robots, allowing them to interact socially with humans. 4. The research community programs interactions with a social robot.

In its final phase, the project aims at using the basic technological, emotional and relational skills acquired in the previous phase to proceed with the conception and programming, through teamwork, of the behaviors of a social/emotional Robot aimed at social interactions. 5 Arrangement and content of the single interventions. Introduction and presentation of the "mission" of the "research community". First coding activities with block programming of Ozobot Construction and programming of a robot Collective Feedback. To achieve student engagement and learning motivation, demonstrations based on research prototypes of social robots will be implemented, as examples of robots that speak and understand the language of human emotions. Demonstrations will also be conducted involving the use of children's mobile phones in class, according to the BYOD (Bring Your Own Device) methodology which involves the use of personal devices at school and which is encouraged by the National Digital School Plan, with the aim of achieving skills through the mediation of new languages and multimedia. The interventions will be based on presentations based on Power-point slides with texts, images, and many videos. They will have an interactive nature, stimulating the interest and learning of students through "problem questions", examples, small exercises and experiential and participatory activities, up to the realization of the final design activity which concludes the laboratory. In detail, a laboratory activity ("hands-on") of an experiential nature on the perception of emotions will be the basis for the realization of the project of a social/emotional robot (project-based learning) and for learning through practice (learning by doing and by creating) of the most basic human-robot social interaction mechanisms. Group work (generally 5 groups of 6 students each) will be organized according to the principles of cooperative learning.

4. Conclusions

Human-robot interactions, in the long run, can have unprecedented developments, largely linked to the intertwining of the multiple co-implicated sciences. As regards pedagogical reflection and didactic planning, the most significant educational effects in this area are linked to the possibility of thinking about human subjectivities, systemically linked to machines that could soon become thinking, as mutant subjectivities, capable of intercepting, thinking and designing from an evolutionary and transformative point of view, not starting from immutable *substances*, but starting from the deep interconnections that evolutionarily link, today more than ever, the natural world and the artificial world, with particular regard to those between humans and robots. In this framework, the proposal is to give rise to a pedagogical planning that knows how to anchor the promotion of a culture of transition and hybridization to the formation of a specific cognitive, cognitive and behavioral equipment that allows the subject to become aware and reflect on one's own identity *mutation* and, at the same time, to develop and act on a decentralized thought and complex causality (Gallelli, 2017 & Pinto Minerva, 2013).

References

Annacontini, G. (2008). Pedagogia e complessità. Pisa: ETS.

Balduzzi, L. (a cura di) (2002). Voci del corpo. Prospettive pedagogiche e didattiche. Firenze: La Nuova Italia.

Barilli, R. (1978). Estetica, quotidianità, tecnologia, in G. M., Bertin (a cura di), Educazione estetica. Firenze: La Nuova Italia.

Barone P. (2014). Embodiment. Formazione e post-umanesimo, in P., Barone, A., Ferrante, & D., Sartori (a cura di), Formazione e post-umanesimo. Sentieri pedagogici nell'età della tecnica. Milano: Cortina.

Bertin, G. M. (1974). L'ideale estetico. Firenze: La Nuova Italia.

Bertin, G. M. (a cura di) (1978). Educazione estetica. Firenze: La Nuova Italia.

Braidotti R. (2014). Postumano. Roma: DeriveApprodi.

Capucci, P. L. (a cura di) (1994). Il corpo tecnologico. L'influenza delle tecnologie sul corpo e sulle sue facoltà. Bologna: Baskerville.

Changeaux, J. P. (1983/1990). L'uomo neuronale. Milano: Feltrinelli.

Damasio, A. (1995). L'errore di Cartesio. Emozione, ragione e cervello umano. Milano: Adelphi.

Decreto Ministeriale 254. 16 november 2012.

Edelman, G., M. (1999). Sulla materia della mente. Milano: Adelphi.

Esposito, R. (2006). La natura umana dopo l'umanesimo, in M., Pireddu, & A., Tursi (a cura di), Post-umano. Relazioni tra uomo e tecnologia nella società delle reti. Milano: Guerini&Associati.

Ferrante, A. P. (2014). Pedagogia e orizzonte post-umanista. Milano: LED Edizioni Universitarie di Lettere, Economia, Diritto.

Frabboni, F., & Pinto Minerva, F. (2014). Una scuola per il Duemila. L'avventura del conoscere tra banchi e mondi ecologici. Palermo: Sellerio.

G.U. n.30. 5 february 2013

Gallelli R. (2017). Processi di soggettivazione, formazione e materialità digitale. In A.Ferrante, J. Orsenigo (a cura di). Dialoghi sul postumano. Pedagogia, filosofia e scienza. Milano: Mimesis.

Gallelli, R. (2011). Insegnare a "imparare con i media". Oltre l'opacità delle tecnologie della conoscenza, in R., Gallelli, & G., Annacontini (a cura di), e.brain. Sfide formative dai "nativi digitali". Milano: FrancoAngeli.

Gallelli, R. (2014). La mente incarnata. Insegnare a comprendere l' unità corpomentale'. In Annacontini, G., Gallelli, R. (a cura di). Formare altre(i)menti. Bari: Progedit.

Gallelli, R. (2017) Processi di soggettivazione, formazione e materialità digitale. In Ferrante, A. & Orsenigo, J. (a cura di). Dialoghi sul postumano. Pedagogia, Filosofia e Scienza. Milano-Udine: Mimesis Edizioni.

Gallelli, R. (2018). Culture del corpo tra Oriente e Occidente. Itinerari formativi. Bari: Progedit.

Greco, A. (2014). Per una pedagogia dell'inclusione. A partire da Vygotskij. Bari: Progedit.

Husserl, E. (1931/2002). Meditazioni cartesiane. Con l'aggiunta dei Discorsi parigini. Milano: Bompiani.

Kelly, K. (1992). Out of control. The new biology of machines, social systems and economic world. Milano: Feltrinelli 1995.

Lapierre, A., & Aucouturier, B. (1978). I contrasti e la scoperta delle nozioni fondamentali. Milano: Sperling&Kupfer.

Le Boulch, J. (1984). Lo sviluppo psicomotorio dalla nascita a sei anni. Roma: Armando.

Longo, G. O. (2006). Il simbionte. Prove di umanità futura. Milano: Booklet.

Longo, G. O. (2014). Post-umano, etica e responsabilità. Riflessioni sistemiche, 10, 7.

Marchesini, R. (2002). Post-human. Verso nuovi modelli di esistenza. Torino: Bollati Boringhieri.

Marchesini, R. (2009). Il tramonto dell'uomo. La prospettiva post-umanista. Bari: Dedalo.

Marchesini, R. (2014). Ibridazioni e processi evolutivi, in P., Barone, A., Ferrante, & D., Sartori (a cura di), Formazione e post-umanesimo. Sentieri pedagogici nell'età della tecnica. Milano: Cortina.

Maturana, H.R., & Varela, F.J. (1985). Autopoiesi e cognizione. La realizzazione del vivente: Padova: Marsilio.

Maturana, H.R., Varela, F.J. (1992). Macchine ed esseri viventi, Roma: Astrolabio-Ubaldini Editore.

Merleau-Ponty M. (1965). Fenomenologia della percezione. Milano: Il Saggiatore.

Merleau-Ponty, M. (1979). Il corpo vissuto. Milano: Il Saggiatore.

Morin, E. (2001). Il metodo I. La natura della natura. Milano: Raffaello Cortina.

Parisi, D. (1999). Mente. Bologna: Il Mulino.

Perla, L. (2013). Per una didattica dell'inclusione. Lecce: Pensa Multimedia.

Perla, L. et al. (2022). La forza mite dell'educazione. Milano: Franco Angeli.

Pievani, T. (2005). Introduzione alla filosofia della biologia. Roma-Bari: Laterza.

Pievani, T. (2008). L'evoluzione della Mente. Le origini biologiche dell'intelligenza, della coscienza, del senso morale. Milano: Sperling & Kupfer.

Pinkler, S. (2002). Come funziona la mente. Milano: Mondadori.

Pinto Minerva, F. (2008). Umano post-umano, per una pedagogia del soggetto mutante, in E. Colicchi (a cura di), Il soggetto nella pedagogia contemporanea. Roma: Carocci.

Pinto Minerva, F. (2013). Prospettive di ecopedagogia. A scuola dalla Natura. In M. Iavarone, P. Malavasi, P. Orefice, F. Pinto Minerva (a cura di), Pedagogia dell'ambiente 2017. Tra sviluppo umano e responsabilità sociale (pp. 173-191). Lecce: Pensa MultiMedia.

Pinto Minerva, F., & Gallelli, R. (2004). Pedagogia e post-umano. Ibridazioni identitarie e frontiere del possibile. Roma: Carocci.

Renna, P. (2021). Salute e formazione. Tra le culture abramitiche del Mediterraneo. Bari: Progedit.

Rivoltella, P.C. (2012). Neurodidattica. Insegnare al cervello che apprende. Milano: Raffaello Cortina.

Rossi, P.G. (2011). Didattica enattiva. Complessità, teorie dell'azione, professionalità docente. Milano: FrancoAngeli.

Seely Brown, J., Collins, A., & Duguid, P. (1989). Situated Cognition and the Culture of Learning, in "Educational Researcher", 1, 18.

Sloterdijk, P. (2001). L'ultima sfera. Breve storia filosofica della globalizzazione. Roma: Carocci.

Spinoza, B. (1677/2006). Etica dimostrata secondo l'ordine geometrico. Torino: Bollati Boringhieri.

Varela F., Thompson, E., & Rosch, E. (1992). La via di mezzo della conoscenza. Milano: Feltrinelli.

Varela, F. (1994). Il reincanto del concreto, in P. L., Capucci (a cura di), Il corpo tecnologico. Bologna: Baskerville.

Varela, F., Maturana, H. (1992). Macchine ed esseri viventi. L'autopoiesi e l'organizzazione biologica. Roma: Astrolabio Ubaldini.

von Bertalanffy, L. (1968). General System Theory. Foundations, Development, Applications. New York: George Braziller.

Vygotskij, L.S. (1925/1972). Psicologia dell'arte. Roma: Editori Riuniti.

Developing skills with educational robotics. Reflections and opportunities from a case study

Federica **Pelizzari**[1], Michele **Marangi**[2] and Simona **Ferrari**[3]

[1] *Catholic University of Sacred Heart, Largo A. Gemelli 1, 20123 Milano, Italy, 0000-0003-2223-7212, federica.pelizzari@unicatt.it*
[2] *Catholic University of Sacred Heart, Largo A. Gemelli 1, 20123 Milano, Italy, 0000-0003-2013-5079, michele.marangi@unicatt.it*
[3] *Catholic University of Sacred Heart, Largo A. Gemelli 1, 20123 Milano, Italy, 0000-0003-3736-1320, simona.ferrari@unicatt.it*

Abstract

This study delved into the contrasting perspectives regarding the role of digital technologies in education, focusing on the potential of educational robotics. It explored the debate between those who advocate delaying children's exposure to digital media and those who highlight the benefits of integrating digital tools into pedagogical practices. Educational robotics emerges as a multifaceted field, rooted in constructivist theories where learning is intertwined with the act of building tangible objects, cultivating a "learn how to learn" mindset. This paper underscored the importance of educational robotics as a 21st-century language, facilitating comprehension of the increasingly interconnected world and communication dynamics. It emphasised that robotics can be introduced to students of all ages, providing opportunities for collaborative learning and individual adaptation.

The study advocated for the cross-curricular integration of robotics in education, focusing on the development of interdisciplinary competencies. It emphasised that knowledge is constructed through active participation, and robotics offers a concrete tool for creative thinking and collaboration.

Furthermore, the research explored the pedagogical potential of educational robotics within the framework of New Media Literacy Education, emphasising the development of soft skills and the empowerment of students in their knowledge journey.

The study's primary research questions revolved around identifying the soft skills that educational robotics and coding can foster in secondary school students. It reported on the conducted educational robotics workshops and their impact on students, highlighting the development of soft skills in alignment with existing literature and increased motivation and relational dimensions.

Keywords
Educational Robotics; Soft Skills Development; Coding; New Media Literacy Education; Coding Paradigms

1. Introduction

The project "Coding&Learning," a collaborative effort between Cremit, ST Foundation, and NGO Acra, presents a compelling case study in the realm of education. This initiative aimed to introduce a training course that leverages coding, educational robotics, and computational thinking as essential tools within a pedagogical and educational framework, designed to enhance the learning experience in secondary school classrooms.

At its core, this project sought to interweave three fundamental dimensions, each integral to its overarching goals.

The first dimension is the media-educational perspective (Rivoltella, 2020), which involves rethinking the relationship with digital technology in the context of modern civic education. In an era defined by digital proliferation, understanding and engaging with digital media are vital components in preparing students for active citizenship (Marangi et al., 2023).

The second dimension focused on the didactic and operational potential of coding, robotics, and computational thinking in direct activities with students (Pelizzari et al., 2023). These tools were not merely seen as being technical skills but as powerful instruments for addressing social and cultural issues. By using these technologies, students could explore and comprehend complex real-world problems and develop the skills to create innovative solutions.

The third dimension of the project was the prospect of developing coding and robotics within a framework of pedagogical co-responsibility (Ferrari & Pelizzari, 2023). This concept acknowledged the importance of involving both formal and informal aspects of teaching to foster the active engagement of learners in their education. It sought to empower students to become protagonists of their learning journey, taking ownership of their educational experiences.

The overarching objective of the "Coding&Learning" project was to generate educational and pedagogical designs that harnessed the potential of coding and robotics. These tools were not viewed solely as a means to acquire technical skills but as a valuable approach in nurturing critical thinking, fostering creativity, enhancing problem-solving abilities, and promoting procedural awareness among students. By combining these dimensions and objectives, the project aimed to create an enriched and holistic educational experience (Koul & Nyar, 2021) that equips students with the skills and mindset needed to thrive in an increasingly complex and digital world.

2. Theoretical Framework

The term "coding" is beset by a pervasive and deeply rooted ambiguity that significantly impedes its consistent and standardised utilisation across diverse educational contexts (Popat & Starkey, 2019; Hennessy et al., 2020). This perplexity also ripples into the realm of "computational thinking," an inseparable companion of coding. To gain a comprehensive understanding of the ramifications of coding within the educational context (Avello et al, 2020; Garcia-Carrillo et al. 2021), it becomes imperative to meticulously articulate our conceptualization of this term, especially when considering its application within the domain of media education (Gretter & Yavad, 2016). The integration of computational thinking, coding, and educational robotics in the educational landscape deepens the learning experience by providing students with a multifaceted approach to problem solving and technology mastery (Bogard et al., 2013; Priemer et al., 2020).

First, computational thinking encourages students to break down complex problems into smaller, manageable parts (Grover & Pea, 2018). It promotes the identification of patterns and regularities within these problems (Hsu et al., 2018), helping students develop a structured approach in finding solutions (Saxena et al., 2020). By exposing students to this kind of abstract problem solving, they gain skills that transcend coding and are highly applicable to a wide range of academic disciplines and professional contexts (Yavad et al., 2017; Pollok et al., 2019). These skills include logical reasoning, algorithmic design, and the ability to analyse and tackle multifaceted challenges effectively (Shute et al., 2017; Palts & Pedaste, 2020; Angeli & Giannakos, 2020).

Coding, as a practical manifestation of computational thinking, allows students to put their abstract problem-solving skills into action (Brennan & Resnik, 2012). As students learn programming languages and write code to create software and applications, they gain a deeper understanding of the relationship between algorithms and their real-world applications (Bers, 2020; Zeng et al., 2023). This practical experience not only enhances their technical competence but also instils the value of perseverance and debugging, as coding often involves trial and error, thus nurturing resilience and adaptability (Saari & Hopkins, 2020).

Educational robotics adds another layer to this didactic framework by making coding and computational thinking tangible (Barth-Cohen et al., 2018). Through the design and programming of robots, students can witness the direct impact of their code on the physical world. This hands-on experience fosters a more profound connection between abstract concepts and their practical applications (Castro et al., 2018; Bozzi et al., 2021). Furthermore, it encourages students to think creatively (Chevalier et al., 2020; Gubenko et al., 2021), as they design and modify robots for specific tasks

(Yang et al., 2020; Misirli & Komis, 2023), and to collaborate as they often work in teams to solve complex challenges (Demetroulis & Wallace, 2021).

The integrated approach of computational thinking, coding, and educational robotics not only equips students with the technical skills demanded by our increasingly digital society but also nurtures a holistic set of competencies (Eguchi, 2014; Ioannou & Makridou, 2018). These include critical thinking, problem-solving abilities, creativity, adaptability, and teamwork, all of which are essential for success in the 21st-century workforce.

To gain a more profound understanding of the role of coding and educational robotics within the domain of media education, it is imperative to explore four distinctive paradigms (Dufva & Dufva, 2016; Dufva, 2018); each shedding light on coding has diverse applications and educational implications:

- Postmodernist Paradigm: This perspective sees coding and educational robotics as a creative endeavour, central to the think-make-improve process. Thriving in informal settings like FabLabs and CoderDojo, it challenges conventional teaching practices. It redefines the relationship between formal and informal learning, emphasising these activities as dynamic and imaginative, fostering innovative problem solving and encouraging students to explore their digital potential beyond traditional boundaries.
- Functionalist Paradigm: In contrast, the functionalist paradigm views coding and educational robotics as a language for enhancing understanding across school subjects. It strongly emphasises integrating programming into the curriculum, bridging abstract concepts with practical application. This approach aims to make complex subjects more accessible and engaging by using coding as a pedagogical aid for subject-specific learning.
- Interpretive Paradigm: This paradigm uses coding and educational robotics to cultivate critical analysis and enhance thinking skills. It encourages students to disassemble problems and understand their underlying structures, fostering creative problem-solving abilities. Emphasising analytical and creative thinking, it focuses on decoding complex challenges and encoding innovative solutions within the context of coding and educational robotics.
- Emancipatory Paradigm: Going beyond traditional education, this paradigm transcends script-based learning. Aligned with political and social movements, it emphasises self-awareness and empowerment. It perceives coding and educational robotics as tools for breaking free from established norms, challenging the traditional educational system, and empowering students to become active agents of change in the digital and broader social context. It encourages critical

thinking about existing power structures and the role of students in shaping the future.

To further explore the implications of these paradigms, two primary axes come into play, shaping our approach to coding education and educational robotics (Ferrari et al., 2017):

Functional Enrolment Axis: This axis views coding and robotics as a tool for adaptation to a technology-driven society. It emphasises the development of critical thinking and a resistance to homogenization. Within this paradigm, coding and robotics is perceived as a means to equip individuals with the skills necessary to thrive in a rapidly evolving technological landscape.

Educational Enrolment Axis: From this perspective, coding and robotics serves as a pedagogical logic that moulds the citizens of tomorrow. It also operates as a social logic, aimed at releasing energy and activating resources for societal progress. In formal contexts, coding and robotics becomes a vehicle for education, fostering the skills and knowledge needed for future citizenship. In nonformal settings, coding and robotics transforms into a medium for personal expression and media activism, encouraging individuals to use technology as a tool for positive change. This axis captures the dual role of coding and robotics as an educational endeavour and a social force for empowerment and innovation.

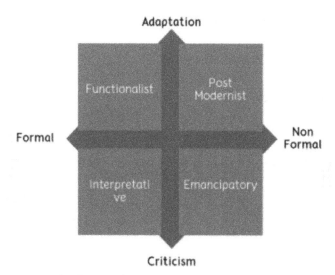

Figure 1: Coding's Paradigms

Transforming the prevailing adaptation perspective among educators is a pivotal undertaking to fully unlock the potential of coding education and educational robotics. This transformation necessitates a renewed emphasis on problem solving, competence development, and the cultivation of creativity within the realm of coding and educational robotics. This holistic approach underscores the importance of nurturing well-rounded individuals who can navigate the intricate web of coding and educational robotics while understanding the broader implications and complexities that it can introduce. Competence, in this sense, extends far beyond the lines of code to encompass the cognitive and socioemotional dimensions of individuals, making them well equipped to excel in a world driven by technology and innovation (Pelizzari et al., 2023).

In this context, coding and educational robotics must be introduced at school as a transversal activity precisely because transversal is the competence it enables to develop (Mertala, 2021). Computational thinking does not need technology; it precedes technology. The adoption of coding and educational robotics as an activity for exemplifying concepts and describing procedures, for solving problems and finding solutions, can be entrusted to teachers of any discipline. This activity, in fact, does not require specific computer skills; it is learnt in an interdisciplinary perspective by mixing creativity and imagination with logic and mathematics.

Learning to be effective must be meaningful, which means that it must motivate and involve pupils in an active way, bringing both logical and creative competence to bear.

3. Research methodology and tools

3.1. The EAS design model

The design model used was conceived in a logic typical of the EAS ("Episodes of Situated Learning" in English) method (Rivoltella, 2013).

> An EAS is a portion of didactic action, that is, the minimum unit of which the teacher's didactic action in context consists; as such it constitutes the barycentre from which the entire didactic edifice is organised. [...] It should be considered as an integral (and integrated) approach to teaching which, certainly, in the case of the use of mobile digital devices finds its preferential application, but which functions regardless of their presence. (Rivoltella, 2013, p. 52)

An EAS is built on three main periods:

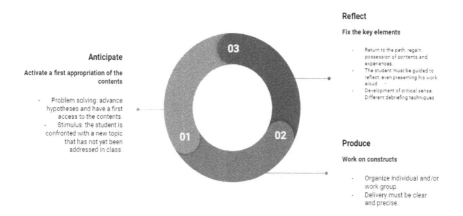

Figure 2: EAS's phases

Working with EAS, also relying on microlearning framework (Hug, 2007), means starting from a careful planning by the teacher (lesson plan) who proposes meaningful situated learning experiences to the students that should end with the realisation of artefacts, thus favouring both the personal appropriation of contents and cooperation and sharing. Operationally, this method identifies a given topic in order to realise a lesson; it is knowledge organised progressively and collaboratively in which the trainer's role becomes that of facilitator and coordinator (Cheng & Szeto, 2016; Warsah et al., 2021).

From this premise, the project was designed by setting four objectives for each class, aimed at fostering the development of critical thinking and collaboration among peer groups (Fung et al., 2016). In addition, an artefact was always put in place to allow pupils to make connections between the various activities, enabling them to create an authentic task, i.e. a teaching activity based on authentic learning that allows pupils in small groups to investigate, discuss, organise or solve problems in real or simulated contexts (Palm, 2009). Each two-hour activity maintained the division between preoperational, operational and restructuring phases typical of EAS, thus enabling the trainer to implement targeted monitoring through specially designed observation checklists (Koth et al., 2009). In addition, the project was equipped with an evaluation rubric (Taggart et al., 1999) to guide the observation checklist and included, in the final phase, the administration of a short self-assessment and peer evaluation (Harrison, 2011) to the students to make them reflect on the progress of the proposed workshop.

From a research point of view, coding and educational robotics in the project phase very effectively lent themselves to the EAS, as the launch phase is by a method managed through motivational problem solving, and the operational phase is by a method involving collaborative and artefact creation. This also made it possible to strengthen critical thinking, which would have been difficult in a project conducted purely in a laboratory or by demonstration/simulation.

3.2. Methods and instruments

This project was guided by two primary research questions that shaped its direction:

- RQ1. Does the implementation of an EAS-based coding lab model bring about a shift in the perspective of coding and robotics, with a particular focus on nurturing soft skills rather than merely emphasising science and computer-related skills?
- RQ2. What soft skills can be cultivated through the introduction of educational robotics and coding in secondary schools?

The research methodology employed a two-fold approach, consisting of parallel and complementary phases (Kothari, 2004).

In the initial phase, a set of online questionnaires was administered to students in the second classes (age 12) of the participating schools, resulting in responses from 153 students (comprising 77 females and 76 males). These questionnaires aimed to explore the students' perceptions of collaboration, utilising various indicators. Additionally, each student received a self-assessment and engaged in peer evaluations at the conclusion of the coding and robotics workshop. Both questionnaires were designed ad hoc based on the research questions and the specific needs of the study. Before distribution, the questionnaires underwent a review process by experts in the field of robotics education and research to ensure their content validity and relevance to the objectives of the study. These experts provided a thorough analysis of the questions, assessing their consistency with existing literature, clarity in wording, and their ability to capture key concepts relevant to the study, thus ensuring validity and relevance to the data obtained.

Simultaneously, in the parallel phase, a separate set of online questionnaires was distributed to the 23 trainers who were actively involved in the project. This group was composed of 11 females and 12 males, ranging in age from 25 to 60. The purpose of these questionnaires was to gain insight into the trainers' perspectives and ascertain whether their attitudes, perceptions regarding coding, and robotics had undergone any notable changes throughout their involvement in the project.

The analysis of the collected data primarily followed a descriptive and exploratory approach (Vittinghoff et al., 2005), aiming to provide a comprehensive overview of the research findings and key observations derived from the responses of both students and trainers. This two-pronged research approach allowed for a holistic exploration of the impact of the EAS-based coding lab model on both students and educators, shedding light on the evolving perspective towards coding, robotics, and soft skills development in secondary schools.

4. Results

The data below represent the answers to the question "Try reading these words. Which two best define group work for you?" Respondents were asked to choose the two words, among those proposed, that best define group work for them; and from the graph, it can be seen that in both initial and final answers, 29% of respondents chose "listening to each other" as one of the two aspects that define group work. Furthermore, the option "proposing one's own ideas" increased slightly from 35% to 36% from the initial to the final phase, also maintaining the most chosen option among the words given, indicating a strong emphasis on the generation of individual ideas in the context of the group. The "role definition" option, on the other hand, decreased from 20% to 18% from the initial to the final phase, which is not surprising, given the emphasis on collaborative and noncooperative management of workshops.

Respondents were asked to choose the teamwork roles they would prefer to take on, and the graph shows that roles such as designer (from 16% to 21%); announcer (from 11% to 15%); materials manager (from 7% to 12%, which included robot management) saw an increase in interest, while roles such as reader (from 11% to 5%) and time controller (from 15% to 7%) saw a decrease in preference at the final stage, thus demonstrating a shift in preferences in teamwork roles over the course of the survey, valuing activation roles over control roles.

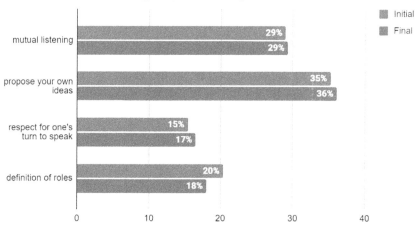

Graph 1: Define group work

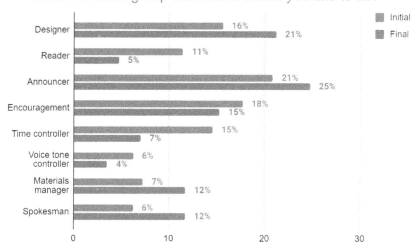

Graph 2: Roles of group work

With regard to the evaluation of self and peers, which was structured to measure the level of agreement or participation in various aspects of group work, it is immediately apparent that the cumulative percentages between "yes" and "more yes than no" refer to very high levels of satisfaction. The data reflect the responses of individuals regarding their experiences and behaviour within the groups. In particular, we note the liking of the activity, where the majority of respondents (52%) said they liked the activity "Very much," suggesting a high level of satisfaction or a positive experience with the task or project.

Furthermore, in the item about listening and respecting group members, the majority (60%) reported that they listened to their groupmates, while a significant percentage (40%) strongly agreed that they respected their turn to speak in group work. This indicates a generally positive environment of respect and attentive behaviour within the group. With respect to contribution in group work, some 55% agreed that they brought their own contribution to the group work. Furthermore, 48% felt they were able to work well in a group, indicating substantial involvement and fruitful cooperation within the group.

Finally, a significant percentage (60%) stated that they complied with the rules of the proposed games, and 56% indicated that they followed the instructions provided in the programme, demonstrating a high level of compliance and adherence to the guidelines established within the activity.

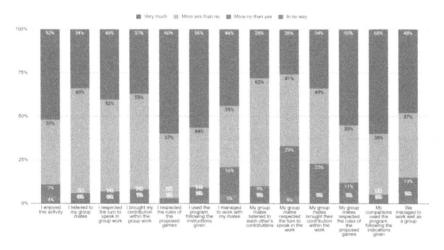

Graph 3: Self and Peer Assessment

As for the trainers, the perceptions of coding and robotics as soft skills activators differ, as can be seen from the following two tables, which show the average response for each item on a Likert scale of 1 to 4: Coding, in comparison to initial expectations, seems to develop collaboration and cooperation skills more (going from 3.13 to 3.40), analysis and critical thinking skills (going from 3.50 to 3.65), and problem-solving and decision-making skills (going from 3.55 to 3.61). Robotics, on the other hand, may be more conducive in problem solving and decision-making (moving from 2.85 to 2.96, still lower than coding), analyse, select and think critically (moving from 2.78 to 2.95); develop creative and unusual but effective strategies (moving from 2.96 to 3.20) and the ability to evaluate and rethink errors (moving from 2.91 to 3.19). In each case, there is no significant change in any of the items, thus demonstrating that the expectations of the trainers were in line with what they then experienced. Finally, it is noted that the standard deviations all fluctuate between 0.61 and 0.93 (for both coding and robotics), demonstrating how the trainers fit into the same range of expectations. Certainly, the trainers point out that they expect less activation of soft skills from robotics than they do from coding.

Table 1
Soft Skills for coding and robotics

	CODING							
How much do you think coding has the potential to facilitate the development of these soft skills?	Solve problems and make decisions	Design, organise and plan	Analyse, select and think critically	Collaborate, cooperate and share	Develop creative and unusual but effective strategies	Simplify complex processes or problems	Evaluate, understand and manage errors	Generalize and transfer what you have learned to other fields
Initial Questionnaire	3,55	3,70	3,50	3,13	3,48	3,48	3,35	3,13
Final Questionnaire	3,61	3,50	3,61	3,40	3,50	3,45	3,40	3,05

	ROBOTICS							
How much do you think educational robotics has the potential to facilitate the development of these soft skills?	Solve problems and make decisions	Design, organise and plan	Analyse, select and think critically	Collaborate, cooperate and share	Develop creative and unusual but effective strategies	Simplify complex processes or problems	Evaluate, understand and manage errors	Generalize and transfer what you have learned to other fields
Initial Questionnaire	2,85	3,13	2,78	3,30	2,96	3,09	2,91	2,70
Final Questionnaire	2,96	2,90	2,95	3,70	3,20	2,95	3,10	2,35

With regard to the dimensions of competence expected and then perceived (expressed on a Likert scale of 1 to 4 and reported as the mean of the responses for each item), however, in coding the expressive, creative and divergent thinking dimensions emerge (which increase from 3.15 to 3.26), as well as the motivational ones (which increase from 3.17 to 3.30). In robotics, on the other hand, those relating to social, relational and affective, empathetic (which rise from 2.30 to 2.70), self-regulation and meta-reflection (from 2.55 to 3) and expressive, creative and divergent thinking (from 2.70 to 3.09) emerge. The pragmatic-motor skills, on the other hand, were equivalent and increased slightly in both coding and robotics. Compared to expectations, they were definitely more amazed by the potential of educational robotics than coding. These latter considerations surprised the researchers greatly because in the scientific literature, the skills developed by coding and robotics are the same, whereas in this research, they are different and complementary to each other.

Table 2
Competence dimensions of coding and robotics

	CODING				
To what extent does coding, in your opinion, have the potential to facilitate the development of these competence dimensions?	Social, relational and affective, empathetic	Self-regulatory and metacognitive	Motivational (involvement, interest, participation)	Expressive, creative and divergent thinking	Praxico-motor
Initial Questionnaire	2,75	3,30	3,17	3,15	2,26
Final Questionnaire	2,78	2,90	3,30	3,26	2,45

	ROBOTICS				
To what extent does robotics, in your opinion, have the potential to facilitate the development of these competence dimensions?	Social, relational and affective, empathetic	Self-regulatory and metacognitive	Motivational (involvement, interest, participation)	Expressive, creative and divergent thinking	Praxico-motor
Initial Questionnaire	2,30	2,55	3,35	2,70	2,50
Final Questionnaire	2,70	3,00	3,30	3,09	2,74

5. Conclusions and future directions

The evolution of the approach to coding and educational robotics from a functionalist perspective to an emancipatory paradigm signifies a profound shift in the way we perceive these educational tools. This transformation is underpinned by a growing awareness that coding and robotics can offer more than just utilitarian skills; they have the potential to be catalysts for empowerment and societal transformation and are not merely technical skills but instruments for empowerment and change. In the functionalist view, coding was often seen as a means to achieve practical objectives or to solve specific problems, primarily within a technical or computational context. While these practical aspects remain important, the emancipatory paradigm widens the horizon.

Within this paradigm, coding and robotics are regarded as tools that transcend the boundaries of technical applications. They become vehicles for fostering critical thinking and encouraging learners to challenge established norms and systems, including the societal and educational status quo. Coding and educational robotics are employed not only to teach students how to write code or operate robots but also to nurture their capacity to think critically, problem-solve creatively, and question the world around them (Dengel & Heuer, 2018). It goes beyond the code itself and considers the broader implications and ethical dimensions of technology (Tsortanidou et al., 2019).

Moreover, this evolving approach to coding and educational robotics goes beyond the traditional confines of STEM (science, technology, engineering and mathematics) education. It finds its home within the broader framework of media literacy education (Rivoltella, 2020), designed to address the complex and dynamic digital landscape that characterises the modern world. It is an educational framework that aims to equip students with the skills necessary, to navigate this media-rich environment effectively (Zecca, 2021).

This expanded perspective recognizes that coding and educational robotics are integral components of media literacy, limiting passive consumption of digital content and stimulating the active creation and critical interpretation of digital media. In this rethought educational landscape, transversal skills, which are necessary to operate in the realms of programming and robotics, are central. The reason for this shift is the recognition that interdisciplinary competencies are invaluable for enhancing students' ability to confront real-world challenges and engage in collaborative, creative and meaningful projects (Leonard et al., 2016).

This evolving approach challenges the conventional demarcation between coding and educational robotics. Rather than treating these as isolated and disparate educational domains, they are now seen as integral components of a unified framework that capitalises on their complementary strengths. This integration recognizes that coding is not a standalone skill but, in fact, an essential element of educational robotics.

In conclusion, the approach to coding and educational robotics is undergoing a significant transformation. It is shifting from a functionalist view to an emancipatory one, expanding its horizons beyond STEM into the realm of media literacy education. This approach values cross-curricular skills and revises the concept of coding as an inseparable part of educational robotics. It will represent a dynamic response to the evolving needs of a digitally connected world, where coding and robotics can empower students and foster critical thinking and creativity.

Acknowledgements

The authors share the conceptualisation and methodological approach of the contribution. Specifically, Federica Pelizzari drafted paragraphs 2, 3 and 4, Michele Marangi wrote paragraph 1, and Simona Ferrari wrote paragraph 5.

References

Angeli, C., & Giannakos, M. (2020). Computational thinking education: Issues and challenges. Computers in human behavior, 105, 106185.

Avello, R., Lavonen, J., & Zapata-Ros, M. (2020). Coding and educational robotics and their relationship with computational and creative thinking. A compressive review. Revista de Educación a Distancia (RED), 20(63).

Barth-Cohen, L. A., Jiang, S., Shen, J., Chen, G., & Eltoukhy, M. (2018). Interpreting and navigating multiple representations for computational thinking in a robotics programming environment. Journal for STEM Education Research, 1, 119-147.

Bers, M. U. (2020). Coding as a playground: Programming and computational thinking in the early childhood classroom. NY: Routledge.

Bogard, T., Liu, M., & Chiang, Y. H. V. (2013). Thresholds of knowledge development in complex problem solving: A multiple-case study of advanced learners' cognitive processes. Educational Technology Research and Development, 61, 465-503.

Bozzi, G., Zecca, L., & Datteri, E. (2021). Interazione bambini-robot: riflessioni teoriche, risultati sperimentali, esperienze. Interazione bambini-robot, 1-448.

Brennan, K., & Resnick, M. (2012, April). New frameworks for studying and assessing the development of computational thinking. In Proceedings of the 2012 annual meeting of the American educational research association, Vancouver, Canada (Vol. 1, p. 25).

Castro, E., Cecchi, F., Salvini, P., Valente, M., Buselli, E., Menichetti, L., Calvani, A., & Dario, P. (2018). Design and impact of a teacher training course, and attitude change concerning educational robotics. International Journal of Social Robotics, 10, 669-685.

Cheng, A. Y., & Szeto, E. (2016). Teacher leadership development and principal facilitation: Novice teachers' perspectives. Teaching and teacher education, 58, 140-148.

Chevalier, M., Giang, C., Piatti, A., & Mondada, F. (2020). Fostering computational thinking through educational robotics: A model for creative computational problem solving. International Journal of STEM Education, 7(1), 1-18.

Demetroulis, E. A., & Wallace, M. (2021). Educational robotics as a tool for the development of collaboration skills. In Handbook of Research on Using Educational Robotics to Facilitate Student Learning (pp. 140-163). IGI Global.

Dengel, A., & Heuer, U. (2018, October). A curriculum of computational thinking as a central idea of information & media literacy. In

Proceedings of the 13th Workshop in Primary and Secondary Computing Education (pp. 1-6).

Dufva, T. (2018). Art education in the post-digital era-Experiential construction of knowledge through creative coding. Aalto University.

Dufva, T., & Dufva, M. (2016). Metaphors of code—Structuring and broadening the discussion on teaching children to code. Thinking Skills and Creativity, 22, 97-110.

Eguchi, A. (2016). Educational robotics as a learning tool for promoting rich environments for active learning (REALs). In Human-computer interaction: Concepts, methodologies, tools, and applications (pp. 740-767). IGI Global.

Ferrari, S., Mangione, G., Rosa, A., & Rivoltella, P. C. (2016). Fare coding per emanciparsi. In P. Limone (a cura di), Modelli pedagogici e pratiche didattiche per la formazione iniziale e in servizio degli insegnanti (pagg. 114-131). Bari:Progedit.

Fung, D. C. L., To, H., & Leung, K. (2016). The influence of collaborative group work on students' development of critical thinking: The teacher's role in facilitating group discussions. Pedagogies: An International Journal, 11(2), 146-166.

García-Carrillo, C., Greca, I. M., & Fernández-Hawrylak, M. (2021). Teacher perspectives on teaching the STEM approach to educational coding and robotics in primary education. Education Sciences, 11(2), 64.

Gretter, S., & Yadav, A. (2016). Computational thinking and media & information literacy: An integrated approach to teaching twenty-first century skills. TechTrends, 60, 510-516.

Grover, S., & Pea, R. (2018). Computational thinking: A competency whose time has come. Computer science education: Perspectives on teaching and learning in school, 19(1), 19-38. 10.5040/9781350057142.ch-003

Gubenko, A., Kirsch, C., Smilek, J. N., Lubart, T., & Houssemand, C. (2021). Educational robotics and robot creativity: An interdisciplinary dialogue. Frontiers in Robotics and AI, 8, 662030.

Harrison, C. (2011). Peer-and self-assessment. Social and emotional aspects of learning, 169-173.

Hennessy, S., Howe, C., Mercer, N., & Vrikki, M. (2020). Coding classroom dialogue: Methodological considerations for researchers. Learning, Culture and Social Interaction, 25, 100404.

Hsu, T. C., Chang, S. C., & Hung, Y. T. (2018). How to learn and how to teach computational thinking: Suggestions based on a review of the literature. Computers & Education, 126, 296-310.

Hug, T. (2007). Didactics of microlearning. Waxmann Verlag.

Ioannou, A., & Makridou, E. (2018). Exploring the potentials of educational robotics in the development of computational thinking: A summary of current research and practical proposals for future work. Education and Information Technologies, 23, 2531-2544.

Koth, C. W., Bradshaw, C. P., & Leaf, P. J. (2009). Teacher observation of classroom adaptation—checklist: Development and factor structure. Measurement and evaluation in counseling and development, 42(1), 15-30.

Kothari, C.R. (2004) Research Methodology: Methods and Techniques. 2nd Edition, New Age International Publishers, New Delhi.

Koul, S., & Nayar, B. (2021). The holistic learning educational ecosystem: A classroom 4.0 perspective. Higher Education Quarterly, 75(1), 98-112.

Leonard J., Buss A., Gamboa R., Mitchell M.S., Fashola O., Hubert T., Almughyirah S. (2016). Using Robotics and Game Design to Enhance Children's Self- Efficacy, STEM Attitudes, and Computational Thinking Skills, "Journal of Science Education and Technology", 25 (6), pp. 860-876.

Marangi, M., Pasta, S., & Rivoltella, P. C. (2023). When digital educational poverty and educational poverty do not coincide: socio-demographic and cultural description, digital skills, educational questions. Q-Times Webmagazine, 15(1), 181-199.

Mertala, P. (2021). The pedagogy of multiliteracies as a code breaker: A suggestion for a transversal approach to computing education in basic education. British Journal of Educational Technology, 52(6), 2227-2241.

Misirli, A., & Komis, V. (2023). Computational thinking in early childhood education: The impact of programming a tangible robot on developing debugging knowledge. Early Childhood Research Quarterly, 65, 139-158.

Palm, T. (2009). Theory of authentic task situations. In Words and worlds (pp. 1-19). Brill.

Palts, T., & Pedaste, M. (2020). A model for developing computational thinking skills. Informatics in Education, 19(1), 113-128.

Pelizzari, F., Marangi, M., & Rivoltella, P. C. (2023). Fare Coding con l'infanzia, una nuova prospettiva educativa. Un'esperienza sulle potenzialità del pensiero computazionale per i bambini di 4 anni. ORIENTAMENTI PEDAGOGICI, 70(2, aprile-maggio-giugno 2023), 69-77.

Pelizzari, F., Marangi, M., Rivoltella, P. C., Peretti, G., Massaro, D., & Villani, D. (2023). Coding and childhood between play and learning: Research on the impact of coding in the learning of 4-year-olds. Research on Education and Media, 15(1), 9-19.

Pollock, L., Mouza, C., Guidry, K. R., & Pusecker, K. (2019, February). Infusing computational thinking across disciplines: Reflections &

lessons learned. In Proceedings of the 50th ACM Technical Symposium on Computer Science Education (pp. 435-441).

Popat, S., & Starkey, L. (2019). Learning to code or coding to learn? A systematic review. Computers & Education, 128, 365-376.

Priemer, B., Eilerts, K., Filler, A., Pinkwart, N., Rösken-Winter, B., Tiemann, R., & Zu Belzen, A. U. (2020). A framework to foster problem-solving in STEM and computing education. Research in Science & Technological Education, 38(1), 105-130.

Rivoltella, P. C. (2020). Nuovi alfabeti. Educazione e culture nella società post-mediale (Vol. 124, pp. 5-220). Brescia: Scholé-Morcelliana.

Rivoltella, P. C., (2013). Fare didattica con gli EAS. Episodi di Apprendimento Situato (pp. 5-241). Brescia: La Scuola.

Saari, E. M., & Hopkins, G. (2020). Computational thinking–Essential and pervasive toolset. Asian Journal of Assessment in Teaching and Learning, 10(1), 23-31.

Saxena, A., Lo, C. K., Hew, K. F., & Wong, G. K. W. (2020). Designing unplugged and plugged activities to cultivate computational thinking: An exploratory study in early childhood education. The Asia-Pacific Education Researcher, 29(1), 55-66.

Shute, V. J., Sun, C., & Asbell-Clarke, J. (2017). Demystifying computational thinking. Educational research review, 22, 142-158.

Taggart, G. L., Phifer, S. J., Nixon, J. A., & Wood, M. (Eds.). (1999). Rubrics: A handbook for construction and use. R&L Education.

Tsortanidou, X., Daradoumis, T., & Barberá, E. (2019). Connecting moments of creativity, computational thinking, collaboration and new media literacy skills. Information and Learning Sciences, 120(11/12), 704-722.

Vittinghoff, E., Shiboski, S. C., Glidden, D. V., & McCulloch, C. E. (2005). Exploratory and descriptive methods. Regression methods in biostatistics: linear, logistic, survival, and repeated measures models, 7-27.

Warsah, I., Morganna, R., Uyun, M., Afandi, M., & Hamengkubuwono, H. (2021). The impact of collaborative learning on learners' critical thinking skills. International Journal of Instruction, 14(2), 443-460.

Yadav, A., Good, J., Voogt, J., & Fisser, P. (2017). Computational thinking as an emerging competence domain. Competence-based vocational and professional education: Bridging the worlds of work and education, 1051-1067.

Yang, Y., Long, Y., Sun, D., Van Aalst, J., & Cheng, S. (2020). Fostering students' creativity via educational robotics: An investigation of teachers' pedagogical practices based on teacher interviews. British journal of educational technology, 51(5), 1826-1842.

Zecca, L. (2021). The Game of Thinking. Interactions between childrens and robots in educational environments. Scaradozzi D., Guasti L., Di

Stasio M., Miotti B., Monteriù A., Blikstein P. (eds.). Makers at School, Educational Robotics and Innovative Learning Environments: Research and Experiences from FabLearn Italy 2019, in the Italian Schools and Beyond, 87-94. Cham (CH): Springer.

Zeng, Y., Yang, W., & Bautista, A. (2023). Computational thinking in early childhood education: Reviewing the literature and redeveloping the three-dimensional framework. Educational Research Review, 100520.

Educational robotics in parents' perceptions

Arianna **Marras**[1], Antioco L. **Zurru**[2] and Antonello **Mura**[3]

[1] *University of Salerno, Giovanni Paolo II Street, 132, 84084, Fisciano, Italy, 0009-0006-2144-3822, amarras@unisa.it, marrasarianna@gmail.com*
[2] *University of Cagliari, Is Mirrionis Street, 1, 09127, Cagliari, Italy, 0000-0001-7783-0393, antiocoluigi.zurru@unica.it*
[3] *University of Cagliari, Is Mirrionis Street, 1, 09127, Cagliari, Italy, 0000-0003-3470-1916, amura@unica.it*

Abstract

The present research investigates parents' perceptions of the effectiveness of extracurricular Educational Robotics (ER) workshops promoted by a school in South Sardinia. The focus of the survey is on parents' ideas with respect to their children's learning processes and skill development, not only technical-practical, but also cross-curricular. Workshop days dedicated to parent-child interaction was scheduled during the ER laboratories, in which students took on the role of tutors to guide parents in programming some educational robots, using different programming languages. The results show that parents were motivated to have their children participate in the robotics project, and had a good perception with respect to the collaboration with their children and with teachers. The distinction according to the educational qualification held by parents appears significant. In summary, the data showed a general parental interest in ER, which was seen as a functional practice for enhancing multiple skills, from social-motivational to STEAM skills.

Keywords

Parents' Perception of Robotics; Educational Robotics; Learning with Robotics; Parenting Relationship; School-Family Collaboration

1. Introduction

In the school context, educational research is progressively focusing its attention on the role that multiple forms of technology can take, including continuous experimentation, to facilitate and stimulate the learning processes, mediated by new digital tools.

With this respect, Educational Robotics (ER) is a specific field of research that, although it does not have an exhaustive definition yet, aims to investigate the use of robots in education to foster learning processes and the acquisition of skills. This configures it as a technology to promote innovative, collaborative and co-participative teaching between educational institutions and families. In fact, parents have a fundamental role as they are in the best position to encourage their children to participate in such activities (Angel-Fernandez & Vincze, 2018), so it is necessary to build an awareness in them that will motivate them to encourage their children to participate and enjoy such projects.

To develop this process, the convictions and dispositions of the various actors involved in various capacities become important.

This work aims to investigate parents' perceptions on the effectiveness of the ER laboratories promoted by the school within an optional program, within the framework of learning processes and the development of skills, not only technical-practical and technological-scientific, but also transversal skills such as, for example, the ability to collaborate. A workshop day dedicated to parent-child interaction was planned at the end of the three-month robotics courses involving the fifth classes of a primary school in South Sardinia during extracurricular hours. During this event, students took on the role of experts and teachers to guide parents in programming some educational robots, using different environments and programming languages. Within the framework just outlined, we administered a survey to find out the perceptions and ideas of parents after the workshop, delving into the topics of ER and various working methods.

2. Theoretical framework

Historically, the relationship between schools and families has always been steeped in tensions and inner contradictions (Deslandes et al., 2015). Starting from the 2000s, the school has gradually implemented forms of attention towards the role that families take on their pupils' education processes, opening up in terms of co-responsibility and solidarity (La Marca, 2005). This relationship becomes authentic and fruitful when the presence of parents at school is characterized by collaborative interaction with teachers and active participation in the planning and organizational dynamics.

Recent body of research show that parental involvement is able to increase families' awareness of and ability to participate in school life (Capperucci et al., 2018). Even in the most critical moments of recent years, the dynamics of collaboration and participation between teachers and families have supported and amplified the learning experience of pupils (Santagati & Barabanti, 2020). Indications in the international arena help to show the extent to which parental engagement and involvement with the school is able to boost pupils' learning development (Povey et al., 2016).

For these reasons, it is deemed necessary to be able to screen parents' ideas about the introduction of digital technologies and specifically ER into the learning process, so that they can be involved in school and teacher decision-making. In an European survey (Eurobarometer, 2015) on the perception of robots, a generally positive idea of robots emerges, even if there is a decrease in the percentage of people expressing a positive opinion compared to the previous 2012 report. Overall, there is an increase in experiences with robots which would appear to be strongly correlated with the attitudes and perceptions of the interviewees towards the use of robots in their daily lives. Although interviewees recognize the advantages of robots, ideas that the same device could steal jobs from humans remain, and that its implementation therefore requires careful management. The same study showed that only 41% of respondents would feel comfortable using a robot at school. The perceptions of ER have already been studied from multiple aspects, both from the perspective of teachers (Bonaiuti et al., 2022) and the perspective of students (Negrini & Giang, 2019). Parents' ideas of ER, on the other hand, do not seem to have been studied extensively yet. Liu (2010) indirectly investigated parents' perceptions of some robotic products, such as programmable bricks, collecting data and finding a positive perception of these devices, which were considered useful as learning tools, albeit with some reservations. Parents play a significant role in their children's lives. Their decision-making and choices can, in fact, become decisive with respect to whether or not their pupils attend robotics courses, purchase a robotic artefact, or use it as a play-learning tool in the family environment. The limited knowledge and experience that parents have on ER could lead to attitudes of non-investment in such tools, not only in terms of home use with their children, but also in relation with the consolidation of beliefs capable of conditioning educational choices and orienting pupils' educational trajectories. Conversely, shared participation in robotics workshop sessions, in which the parent-child dyad works together, could improve their interaction (Bers, 2007). Several papers analyse dyadic family collaboration in the context of robotics courses (Bers & Urrea, 2000; Beals & Bers, 2006; Cuellar et al., 2013; Roque et al., 2016; Zabik & Tanas, 2018), opening interesting research scenarios for further investigations.

3. Methods and instruments

The present work takes the form of an exploratory and pilot research with the objective of doing a preliminary survey with a small sample and being able to verify, from the parents' point of view, the multiple aspects involved in education with robotics. This exploratory research takes the form of a survey study at the end of ER workshop activities.

This made it possible to overcome any "impasses" arising from a negative perception of ER practice, anchored, for example, in a superficial knowledge of the topic or a still limited experience of the practice.

The ER labs were held on Saturday mornings during extracurricular hours. In each lab session, students together with teachers learned how to program different robots in teams through a learn by doing approach. Students worked with block-based programming languages. During the 20-hour lab session (Figure 1), students increased their expertise in ER and learned how to write more complex programs that involved the use of the software's multiple functionalities. The lab sessions were organized as follows: at first, the teacher provided the materials and the goal to be achieved, then the students worked on the code to achieve the goal or implement a piece of code, concluding with a debriefing phase, to show the whole class the work of the individual groups and to reflect on the results. The last part of the workshop also involved the parents of the pupils working with their children, in pairs or small groups. The children took on the role of tutoring, guiding the parents in programming the robots.

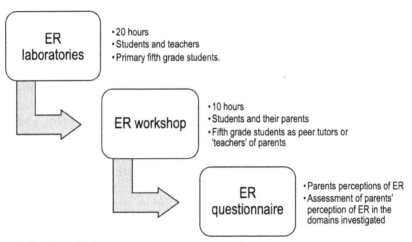

Figure 1: Project design

The survey consisted of 45 items, some in the form of a 5-point Likert scale (from 1 = totally disagree to 5 = totally agree), formulated to collect parents' attitudes towards certain aspects, with some dichotomous questions, purposely designed by the authors. A preliminary section of the questionnaire collected parents' biographical information (gender, age, occupation, educational qualification). The educational qualification question consisted of a multiple-choice option: "middle school license", "secondary school diploma", "bachelor's degree", "master's degree" or "phD".

The questionnaire was shared with fifty parents of fifth-grade pupils involved in the national operational programme "coding and educational robotics". Thirty-seven valid questionnaires (74% of parents involved) completed by parents were returned (age: mean=46.05; median=47.5; ds=5.84; min=29; max=62). In order to assess parents' perceptions on the effectiveness of ER and the imprint this practice can leave on their children in terms of orientation, several aspects were assessed, from progress in socialization dynamics to learning dynamics, motivation and participation. Specifically, the questionnaire investigated the acquisition of STEAM (e.g., "In my opinion, after the educational robotics experience, my son/daughter has improved his/ her approach to STEAM disciplines"), socio-relational (e.g., "Since the educational robotics experience began, my son/daughter has built long-lasting friendship networks"), affective-emotional (e.g., "Since the educational robotics experience began my son/daughter gives up more often when faced with complex situations") and executive function building skills (e.g., "In my opinion, after the educational robotics experience, my son/daughter has become more adept at finding a solution to problems"). Aspects related to the motivational sphere of pupils and parents (e.g., "I thought the ER laboratory was a good opportunity to see how my son/daughter works"), collaboration measured both as collaboration with one's own children in the act of planning (e.g., "Cooperation with my son/daughter made me feel good"), and with teachers in the organization and realization of the final artefact (e.g., "The interaction with the teachers of the project was complicated"), while also exploring the perception of ER as an orienting and useful device in the extracurricular sphere were also studied (e.g., "I believe that this experience helped to orient my son/daughter in the choice of secondary school" or "After this experience my son/daughter started to build things independently").

4. Data analysis and results

Data analysis was performed using the software *Jamovi*[2]. Descriptive statistics (Table 1) and the nonparametric Kruskal-Wallis test were done to compare the data, assessing in what terms there were significant differences in parents' perceptions. Descriptive-level dimensions were calculated by evaluating the mean value.

---

[2] The Jamovi project. Jamovi (Version 2.3). (2022).

Table 1

Descriptive statistics in relation to the aspects investigated

Aspects investigated	Mean	Median	SD	N^ Item
Affective-emotional skills	3.75	3.83	0.451	8
Executive functioning	4.04	4.00	0.477	6
Formative orientation	3.46	3.50	0.794	2
Social-relational skills	3.85	3.83	0.564	6
STEAM skills	3.74	3.83	0.451	6
Student motivation	4.20	4.17	0.492	6
Parental motivation	4.14	4.25	0.625	4
Perception of parent-child-teacher collaboration	4.50	4.50	0.612	2
Perception usefulness of ER in daily life	3.90	4.00	0.661	5

Table 2

Kruskal-Wallis test to assess the differences between the medians of the answers given by the subjects in relation to their educational qualifications

Aspects investigated	χ^2	gdl	p	ε^2
Affective-emotional skills	12.68	3	.005 **	0.373
Executive functioning	9.77	3	.021 *	0.287
Formative orientation	8.25	3	.041 *	0.243
Social-relational skills	8.09	3	.044 *	0.238
STEAM skills	4.82	3	.185	0.142
Student motivation	5.09	3	.165	0.150
Parental motivation	6.89	3	.076	0.203
Perception of parent-child-teacher collaboration	5.45	3	.142	0.160
Perception usefulness of ER in daily life	4.47	3	.215	0.132

The distinction according to the educational qualification possessed, for example, appears significant. Parents with lower educational qualifications (middle and high school diploma) are more likely to believe that ER can foster the development of their children's affective-emotional, social-relational skills, and executive functioning with respect to parents with a college degree. In contrast, parents with higher education degrees are more likely to think of ER activities as useful in guiding their children in choosing school and future career paths. We can observe good motivation of the parents to have their children participate in the robotics project is observed (m = 4.14; sd = 0.625). Parents report the enthusiasm of pupils when talking to their children about the course. The most encouraging finding shows an optimal perception of parents with respect to collaboration with their children and teachers (m = 4.5; sd = 0.612), supporting a fruitful interaction even in terms of parent-child relationship.

Among the most stressed aspects according to parents' perceptions are the affective-emotional dimension (p=.005), the socio-relational dimension (p=.044) and the development of cognitive functions (p=.021). The differences are more significant among parents with a medium and high level of education (high school graduates and university graduates). The parents' opinion of ER as a tool for orientation in future school choices also appears to be statistically significant (p=.041). Furthermore, it should be emphasized that a higher educational level of the parents does not also correspond to a higher motivation of the parents with respect to participation in the project (p=.060).

Table 3
Descriptive statistics in relation to the Parents Perception usefulness of ER in daily life

Items	Mean	Median	SD
1. After this experience, my son/daughter started programming at home (e.g. playing in code.org or in Scratch)	3.68	4	0.915
2. After this experience my son/daughter started to build things independently.	3.65	4	0.857
3. After this experience my son/daughter has understood how different devices (software) work.	4.32	5	0.818
4. After this experience my son/daughter has learnt about the various parts that make up different robotic devices (hardware)	4.35	5	0.789
5. After this experience my son/daughter is able to build an electronic device.	3.51	3	0.901

However, aspects related to the motivational dimension of the children and education in STEAM disciplines are considered relevant by the whole group of parents: after the course, according to the parents' statements (Table 3), the pupils mostly understood the functioning of software (m=4.32; sd =0.818) and hardware (m=4.35; sd =0.789), confirming the idea that ER can enhance the students' STEM competences. Parents admitted that, thanks to the ER laboratory, their children started programming (m=3.68; sd=0.915) and building objects at home (m=3.65; sd=0.857).

5. Conclusions and future prospects

In general, data shows that parents are interested in ER, and that is seen as a functional practice for enhancing various skills, from social-motivational to STEAM skills. This is a preliminary analysis, conducted with a specially constructed tool to understand how parents perceive the interventions and how they evaluated their collaboration with the different actors involved in the educational process. As the results show, the most stressed aspects such as affective-emotional and socio-relational dimensions are very interesting to explore; in fact, in our study a statistically significant difference can be seen depending on the educational level of the parents involved. At the qualitative level, some research shows these aspects as preeminent in the general perception of parents with respect to pupils' school experience (Mura et al., 2020; Zurru et al, 2021). Lin et al. (2012) following their study on parents' perception of the use of educational robots, revealed that most parents would consider robots useful for their children, confirming our observations, for example for what concerns STEM education. However, parents felt unsafe when teaching their children to use robots. We imagine it could be useful, in order to increase parents' self-efficacy in implementing these practices, to be able to co-design and implement robotics laboratories that involve parents with their children in a sufficient amount of time. The parents' high motivation to participate in ER activities (m=4.14; (sd=0.625), the positive perception of collaboration with their children and teachers (m=4.50; sd=0.612) and the idea that ER workshops are useful for the acquisition of daily skills (m=3.90; sd=0.661) show parents' proactivity towards ER. In agreement with Toh et al. (2016), also our results show that it would also be appropriate to involve parents and educators in educational programs and robotics laboratories: this would allow them to increase their chances of success. Moreover, the lack of support from parents would not only limit educational robots to applications within the classroom, but would limit the expansion of their potential in different (e.g., informal) settings in order to consolidate interaction and the parent-child educational relationship. These ideas focus on ER and do not include robots with social characteristics and capable of communicating with the environment and people. Regarding this type of robot Lee et al. (2008) showed that although students and parents thought more about their use in the school setting than teachers, no parent would want robots to replace teachers.

In the future, we believe it may be interesting to compare the data with the perceptions of teachers and pupils and, with a larger sample, to proceed to a factorial analysis to identify the dimensions that the various items have in common from a statistical point of view and to be able to assess their possible educational implications in terms of educational orientation. Future research, based on a project involving a complete robotics course shared between parents and children, and thus with a greater number of observations in different situations, would allow research to deepen what we have observed so far. Setting up a permanent robotics workshop for parents and children would allow us to understand the nature of parents' beliefs and foster trust in robotics teaching, effectively contributing to the development of parental relations and a broader educational partnership. In this direction, functional collaborations between the school and family could be verified for the implementation of co-participated educational projects that see the multiple realities of pupils working towards an increasingly aware education.

Ultimately, it is necessary to remember that in structuring the teacher's professional profile, the ability to involve parents in workshop activities acquires significant importance. On the one hand, it allows the teacher to leverage their soft skills (e.g., collaboration) and on the other hand, parents have the possibility to authentically gain a clear awareness of learning processes. This approach can help the emergence of a truly shared teaching-learning process that places the student and the learning goals in the bull's eye, while the agencies that collaborate to make this happen successfully right around them.

Authors' contribution statement

The article is the result of the work of all authors who share the conceptual and methodological approach of the research. In particular, Arianna Marras wrote paragraphs 1, 2, 3, and 4. Paragraph 5 is equally attributed to all authors. Antonello Mura coordinated the research work. Antioco Luigi Zurru coordinated and supervised the writing of the contribution.

Acknowledgements

We would like to thank the head of the school who participated in the survey, teachers, ER experts, students and their parents for enthusiastically accepting the proposal to participate in this research.

References

Beals, L. and Bers, M. (2006). Robotic technologies: when parents put their learning ahead of their child's. Journal of Interactive Learning Research, 17(4), pp.341-366.

Bers, M. U. (2007). Project interactions: a multigenerational robotic learning environment. Journal of Science Education and Technology, 16, 6, 537–552.

Bers, M.U. and Urrea, C. (2000). Con-science: Parents and children exploring robotics and values. A. Druin.

Bonaiuti, G., Zurru, A. L., & Marras, A. (2022). La robotica educativa nelle percezioni degli insegnanti. RicercAzione, 14(1), 131–156.

Capperucci, D., Ciucci, E., & Baroncelli, A. (2018). Relazione scuola-famiglia: alleanza e corresponsabilità educativa. Rivista Italiana di Educazione Familiare, 2028, n.2, 231-253.

Cuellar, F., Penaloza, C., Garret, P., Olivo, D., Mejia, M., Valdez, N. and Mija, A. (2014). Robotics education initiative for analyzing learning and child-parent interaction. In Frontiers in Education Conference (FIE), 2014 IEEE (pp. 1-6). IEEE.

Deslandes, R., Barma, S., & Morin, L. (2015). Understanding Complex Relationships between Teachers and Parents. International Journal about Parents in Education, 9(1).

Eurobarometer. (2015). Special Eurobarometer 427 Autonomous systems. European Commission.

La Marca, A. (2005). Famiglia e scuola. Armando Editore.

Lee E., Y. Lee, B. Kye, and B. Ko, "Elementary and middle school teachers', students' and parents' perception of robot-aided education in Korea," in Proc. World Conf. Educ. Media Technol., Vienna, Austria, 2008, pp. 175–183.

Lin, C. H., Liu, E. Z. F., & Huang, Y. Y. (2012). Exploring parents' perceptions towards educational robots: Gender and socioeconomic differences. British Journal of Educational Technology, 43(1), E31-E34.

Liu, E. Z. F. (2010). Early adolescents' perceptions of educational robots and learning of robotics. British Journal of Educational Technology, 41, 3, E44–E47.

Mura, A., Zurru, A. L., & Tatulli, I. (2020). Inclusione e collaborazione a scuola: un'occasione per insegnanti e famiglia. Italian Journal of Special Education for Inclusion, VIII (1), 260–273.

Negrini, L., & Giang, C. (2019). How do pupils perceive educational robotics as a tool to improve their 21st century skills? Journal of E-Learning and Knowledge Society, 15(2), 77–87.

Povey, J., Campbell, A. K., Willis, L. D., Haynes, M., Western, M., Bennett, S., Antrobus, E., & Pedde, C. (2016). Engaging parents in schools and building parent-school partnerships: The role of school and parent organisation leadership. International Journal of Educational Research, 79, 128–141.

Santagati, M., & Barabanti, P. (2020). (Dis)connessi? Alunni, genitori e insegnanti di fronte all'emergenza Covid-19. Media Education, 11(2), 109–125.

Vel Żabik, K. P., & Tanaś, Ł. (2018). Shared cooperative activities in parent-child dyads in an educational robotics workshop. Proceedings of the 15th International Conference on Cognition and Exploratory Learning in the Digital Age, CELDA 2018, 194–200.

Zurru, A. L., Tatulli, I., Bullegas, D., & Mura, A. (2021). DaD: tutti
bocciati? Professionalità docente e rapporti di cura nell'esperienza dei
genitori. Annali Online Della Didattica e Della Formazione Docente,
13(22), 136–151.

Words and images in action: the social robot Nao as a knowledge mediator

Giacomo **Antonello**[1], Lucrezia **Bano**[2], Sandro **Brignone**[3], Renato **Grimaldi**[4] and Silvia **Palmieri**[5]

[1] *Dept. of Philosophy and Education Sciences, University of Turin, Italy, 0009-0004-2823-7883, giacomo.antonello@unito.it*
[2] *Dept. of Philosophy and Education Sciences, University of Turin, Italy, 0009-0001-8715-6491, lucrezia.bano@unito.it*
[3] *Dept. of Philosophy and Education Sciences, University of Turin, Italy, 0000-0001-7266-5405, sandro.brignone@unito.it*
[4] *Dept. of Philosophy and Education Sciences, University of Turin, Italy, 0000-0002-9794-1821, renato.grimaldi@unito.it*
[5] *Dept. of Philosophy and Education Sciences, University of Turin, Italy, 0000-0001-5123-4761, silvia.palmieri@unito.it*

Abstract

This work presents an aspect of interaction between children and robots, focusing on how the social robot Nao can "embody" concepts by activating words and images, in pursuit of "embodied semantics". Teachers or educators construct a knowledge base that children explore by interacting with Nao, learning through a playful-educational process that incorporates multiple languages. Thus, the social robot becomes a tool for mediating knowledge. The applications can be adapted to various disciplines and cultural contexts, supporting the educational needs of users. The experience unfolds in two phases. The first involves training the robot: the educator creates a knowledge base, presenting and having Nao memorize cards containing images or words, and then enriching them with metadata. In the second phase, children directly interact with the social robot using the previously prepared cards. Through artificial vision and dynamic communication, Nao responds, elucidates the meaning, and "brings to life" the displayed concepts, facilitating their learning.

Keywords
Social Robot, Child-Robot Interaction, Embodied Semantic, Knowledge Base, School and Education

1. Social Robots in Education

In the past fifteen to twenty years, social robots have been experimentally employed in various situations and contexts (Bartneck *et al.*, 2020; Korn, 2019; Bhaumik, 2018; Breazeal, 2016), including fields such as healthcare, tourism and commercial spaces, and in public and private offices, among others. In the education sector, numerous research studies and applications have been developed, showing how these systems have the potential to progressively become part of the educational and training infrastructure (Alnajjar *et al*, 2022; Belpaeme, Tanaka, 2021; Belpaeme *et al.*, 2018; Mubin *et al.*, 2013).

Recent scientific literature defines a social robot as an artificial intelligence (AI) with a body, a physical agent with a degree of autonomy, designed and programmed to communicate and interact with humans in a natural way, following socially shared norms (Sarrica *et al.*, 2020). In other words, a social robot should adopt behaviors similar to those we use when interacting with other people (Belpaeme, 2022). The purpose of these machines is to assist people in achieving positive outcomes in various life situations or, more generally, to pursue the goals that society sets for its development (Breazeal *et al.*, 2016). As mentioned, the usage scenarios are broad and diversified, many still being explored and experimented with, others more established (Grimaldi, 2022; Youssef *et al.*, 2022).

Particularly in the field of education and training, social robots have been used, albeit marginally, as "robots" in a "passive" role to study specific disciplines like science, technology, engineering, and mathematics (STEM), and more significantly, as "social robots", serving as pedagogical agents providing educational experiences to students through social interaction (Lehmann, Rossi, 2019). In this latter case, social robots are active partners in the learning environment and transmit its contents. Compared to virtual agents, which are software that helps acquire certain knowledge and skills through intelligent algorithms (the so-called Intelligent Tutoring System, ITS), social robots offer a more natural and engaging interaction experience, with a physical sharing of spaces and objects, as well as social behaviors that stimulate learning. This approach is referred to as ITR, Intelligent Tutoring Robot (Yang, Zhang, 2019).

Much research has been conducted with children (early childhood and primary education), partly with middle and high school students, and to a lesser extent with university students and adults. The interventions mostly focus on metacognitive and affective outcomes (such as the level of attention on the task demonstrated by the student, positive emotions expressed in learning, responses and questions posed to the robot, etc.), and secondarily on cognitive performance (level of acquired knowledge, application of learned notions, synthesis, etc.) (Belpaeme *et al.*, 2018). For example, social robots have been experimented with in primary and secondary schools for teaching mathematics, first and second languages, or other curricular subjects, handwriting, as well as for developing cognitive-relational skills in children with specific learning disorders or disabilities, etc. (Woo *et al.*, 2021).

During the educational interventions carried out, social robots mostly played the role of "tutors" and "teachers" supervising learning; to a lesser extent, they have been experimented as "study companions" in a peer-to-peer role, or as "novices/beginners", where the student themselves teach the robot the notions to be learned (Belpaeme *et al.*, 2018). Currently, humanoid robots are used in restricted scenarios and with a standard format: well-defined, short lessons on specific topics, with relatively limited flexibility regarding the curriculum and limited adaptation to individual demands in the class context. However, it is possible to work with a higher level of personalization in moments of interaction with an individual student or in a small group.

Based on these experiences, recent studies have also explored the possibility of using social robots for cognitive purposes, such as evaluating the knowledge and skills possessed by children in a curricular educational context. In this regard, applications on social robots have been created for administering tests or questionnaires that measure the learnings and competencies acquired by individual children within the class group (Abbasi *et al.*, 2022; Varrasi *et al.*, 2018).

To date, there are numerous technical challenges that social robots must and will need to address to become effective "educational companions" for humans. Primarily, these include perception and movement within a physical context and, even more importantly, the correct understanding of the social world in which these systems operate. This understanding encompasses voice recognition, processing of verbal and non-verbal social signals, adaptation of educational content, and more. Additionally, ethical issues and privacy concerns loom on the horizon—equally significant in the development of a positive relationship between robots and humans, aimed at supporting the latter. In this particular case, the intent is to illustrate an example of how the capabilities of an agent, equipped with AI and a physical body (considering its limitations), can be leveraged within the educational and school environment.

To guide the reader through this work, it is emphasized that the research adheres to formal procedures through a logical method, wherein knowledge is fully embedded within the assumptions of the model itself, and development is top-down, from theory to facts (data). Here, validation is local: the model's engine is software embedded within a social robot, and thus, correct procedure execution provides "local proof" of validation (Ricolfi, 1996). Moreover, the paper outlines a proposal for activities that can be performed by an Intelligent Tutoring Robot (ITR); potential applications and reflective insights aim at developing as effective and personalized interactions between children and robots as possible, always under the teacher's supervision. The goal is to enhance and make the students' learning experience more engaging, as well as to support the educational efforts of teachers within the school context.

2. The Social Robot as a Mediator of Meanings and Knowledge

Within the framework just outlined, the subsequent project is introduced, which involves the use of *SoftBank Robotics*' social robot *Nao* in a didactic unit within a primary school class (see sect. 3). The robot conveys concepts – expressed in linguistic and figurative form – through its words and body actions, in the direction of an "embodied semantics". This concept emerges from the intersection of several disciplines, particularly linguistics, cognitive psychology, and neuroscience (Meteyard *et al.*, 2012); it is also part of a broader movement in cognitive science, known as "embodied cognition". Embodied semantics posits that understanding language is not just an abstract mental process, but is closely connected to our sensory and motor experiences. In this perspective, our ability to make sense of words is influenced by the way our bodies interact with the world, in particular contexts and situations. For example, when processing words that refer to movement, such as "running", certain parts of the brain responsible for physical movement can be activated.

These reflections lead to interesting connections with the field of social robots and their applications in an educational scenario. Indeed, social robots – just like people – are equipped with a physical body and are programmed to interact in specific social contexts, often mimicking human behaviors, facial expressions, and gestures. In this sense, such systems can facilitate communication, understanding, and learning of certain concepts, for example, by accompanying words with gestures or other actions to make the message more natural and closer to human experience. In this regard, it can be said that at least three characteristics possessed by social robots can connect and lead them in the direction of "embodied semantics" (Guggemos *et al.*, 2022; Lugrin *et al.*, 2022).

The first aspect, which is evident, is the human encounter with a physical presence in a room, a body in space. It's not just about a voice or graphic interface, a virtual AI that – so to speak – is "elsewhere" and communicates with us beyond a screen or a static speaker. It's something more: in educational contexts, the social robot has a body that has a biology-inspired appearance, with more or less anthropomorphic or zoomorphic features. The traits are simple and friendly, often "childlike and sweet", positively influencing people's perception and expectations towards them.

A second element relates to the social capabilities of the robot and how it establishes a relationship with the human being. Social robots communicate verbally and non-verbally, using a combination of natural language, movements, lights, sounds. Eye contact, voice characteristics, and exhibited gestures can make the interaction experience, and in our case, the transmission of content and the educational experience of children at school, more engaging and meaningful. The detection of emotions and the adaptation of behavior can then increase the sense of empathy.

A third characteristic has to do with the degree of intelligence and autonomy manifested, that is, how far the social robot is able to perceive the characteristics of the environment in which it finds itself and the potentially present people (with its sensors), to plan actions to be carried forward in that particular context (algorithms and AI processes), and then act on it in an attempt to achieve specific goals (with motors and actuators); all this with more or less supervision by external subjects.

Taking these reflections into an educational context, having a social robot that shares some tasks with the teacher can be useful. In the case of our work, the machine can indeed become a mediator of content – that "sets in motion" and conveys knowledge – giving meaning to concepts expressed through words and images. Thus, the teacher plays the role of guide, of preparation and organization of content, and of programming the robot. The latter is delegated with more specific tasks, which can meet needs for personalization, individualized or small group learning, and management of interactions among students during activities in an inclusive perspective (Zecca, Datteri, 2023; Marchetti, Massaro, 2023) (see sect. 3).

The prototype of interaction between children and robot was developed within the activities of the *Laboratorio di Simulazione del comportamento e robotica educativa "Luciano Gallino"*, founded in 2018, thanks to the Project of Excellence of the Department of Philosophy and Education Sciences (University of Turin). The Laboratory is engaged in theoretical-practical research in educational and care contexts, proposing innovative training and experimental teaching at both university and school levels, and with local entities and companies, aiming to train personnel, teachers, and students for a conscious use of technology.

3. The development of the activity

The project was conceived to be divided and carried out in two main stages. The first involves the training phase of the social robot, in which educational content is - so to speak – "taught" to the system. Teachers or educators create a series of cards with images referring to the contents they intend to develop with the class or a group of children. These can contain words, places, historical figures, events, diagrams, figures, or any content of interest for a certain educational topic. Subsequently, the cards are shown, one at a time and in succession, in front of the robot's camera (which takes a photo of each). For each card, it is necessary to input into the system the information intended to be associated: the images are thus "labeled" and linked to identified and descriptive metadata. Using computer vision techniques, it's possible to segment a precise area of the captured frame to which the robot should attribute relevance and connect the related contents. For example: the word "recite" in English, associated with the corresponding label "recite", or "recitare" in Italian, to which the interpretation of a certain passage or poem by the robot in subsequent interaction phases can also be concatenated (Figure 1). The process leverages the functionalities offered by the *Choregraphe* platform (proprietary software of SoftBank, that manages the social robot Nao). This choice seemed closer to the potential needs of teachers and educators, including those who are attempting programming for the first time.

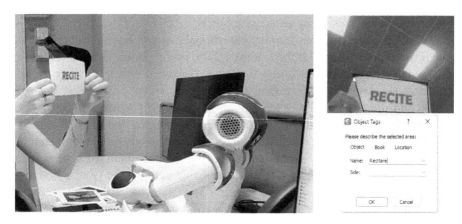

Figure 1: Nao's training phase on the Choregraphe platform. A training-text sheet is displayed in front of the robot's frontal camera, so that the area with the content of interest can be selected. The relevant information (metadata) is then associated with it (photo: S. Palmieri)

The second phase is concerned with the interaction between the social robot and the children in the classroom, where the latter can engage with the content previously developed by the teacher. Students, individually or in small groups, have access to the images related to the educational unit under consideration and can initiate a conversation with Nao. During the dialogue, they show the robot the cards with the images and through it, learn and delve into the content associated with these images (Figure 2). The system is capable of decoding and understanding the student's natural language and recognizing the patterns present in the images, previously learned. In turn, Nao's responses are provided in verbal and non-verbal language, thus giving a coherent meaning – a "body", with words, intonation, and gestures – to the concepts developed in the cards.

In particular, the robot's outputs highlight three main characteristics. Firstly, Nao can repeat responses, provided to a given image or stimulus presented to it, an unlimited number of times. In a sense, what might be termed the "patience" of a robot is higher than that of a human, as it never tires. Repetition is crucial for providing children (each with their own peculiarities) the freedom and time necessary to learn.

Moreover, the responses can be replicated for any user and remain identical over time, or within certain ranges of variation designed by the teacher, also considering the educational needs of the children in the class. This second aspect allows students to interact with an agent that, at least in its potential, is free from biases that might unintentionally be present in a human teacher. Additionally, since the robot's behavior is relatively consistent over time and predictable in its responses, this facilitates interaction and learning by children with behaviors associated with the autism spectrum (Saleh *et al.*, 2021, Pennisi *et al.*, 2016).

The third characteristic that emerges is related to conversation rules. The dialogue with the robot, programmed to follow certain criteria and sequences of communicative exchange, fosters the internalization and reinforcement of correct conversation norms by the children. This leads to a respectful confrontation with others and pays attention to shared criteria of interaction. For example, using certain expressions to question Nao or respecting turns in conversation represent, on one hand, a way to meet the inherent needs of the robot's system, but at the same time, such rules can be transferred to the context of the class group with classmates. Nao, just like "another peer", shows to have certain needs for understanding communication, which the community must respect in order to include him. Therefore, this social logic lays some important foundations for the transmission of relational skills.

In summary, following the process just described, teachers or educators can manually construct and assemble the content of the database and, at the same time, realize the interactions that will take place between children and robot. The formulation of input questions, output responses, and the robot's gestures can be structured in the way the adult finds clearer and more inclusive for the child, for example giving more or less space and shape to the communication with the student. In this sense, the research group has created a video that exemplifies some possibilities of the designed model, visible at the following link: *www.youtube.com/watch?v=-vJDU_PmN04*. The prototype can obviously be adapted to different contents and levels of education.

Figure 2: The picture on the left shows some examples of educational cards created, containing images and words. On the right, researcher S. Palmieri shows a card to the social robot for its recognition and explanation (photo L. Bano)

4. Applications and possible developments

Applications in the educational field are particularly aimed at students who struggle with traditional teaching methods, with inclusive approaches and laboratory enhancement activities that integrate frontal lessons and curriculum subjects. Nao, for example, can interact with children in the early grades of primary school to introduce them to writing or help them consolidate the relationship between graphemes and meanings in their mother tongue. Interaction with the robot can also reinforce graphomotor skills such as eye-hand coordination and fine motor skills (Tapus *et al.* 2007; So *et al.* 2018). Empirical-semantic reinforcement can also occur with older students in situations of learning a foreign language, as an approach for acquiring "basic" vocabulary connected to social moments (greetings, introducing oneself, etc.). With the assistance of competent staff, such activities can also be implemented in the context of a sign language laboratory, where, for example, the learner learns the meaning of a gesture represented graphically and performed by the robot. Interaction with the social robot then enhances the social and relational skills of the class group, also from an inclusive perspective. In fact, the system can convey social stories, a sort of "pill" that shows students some positive and correct models of interaction with others (Horner *et al.*, 2005).

Another horizon is represented by extra-curricular realities, such as museums, archaeological sites, parks, and art galleries, where Nao could act as a guide within dedicated educational paths for the enjoyment of content and activities available, with a special focus on non-verbal communication. Moreover, one can envision a module on civic and road education, where images are related to road signs and other urban furnishings, and the robot's outputs, through speech and gesture, convey meanings such as "stop", "slow down", "go ahead". Similarly, a music workshop is envisaged where, in front of a score, Nao pronounces the names of the notes, reproduces their tonality, or keeps time with the clapping of its hands.

Finally, a machine learning workshop can be structured for secondary school students, where they undertake both phases of the activities described in this paper. Boys and girls are guided in creating the knowledge base, selecting images, instructing the robot, and finally, verifying the correctness of the process. In this way, in addition to firsthand discovering how supervised learning works from a technical standpoint (also with more performant machines and through big data), boys and girls come into contact with the world of AI, understanding how some of its processes can be influenced by human choices. This type of competence is crucial for education towards active citizenship: training individuals-citizens who, in the near future, will be required to have a knowledgeable and critical ability to use certain tools that will increasingly belong to their everyday life context, such as communicating with voice chatbots (e.g., from Alexa or ChatGPT and their successors), interacting with future social robots or other AI-equipped devices, querying complex databases, etc.

5. A few conclusions

At this historical moment, when artificial intelligence and neural networks enjoy great popularity and use, we find ourselves in front of machine learning systems whose functioning can sometimes be "cryptic" even for sector experts (hence the talk of "black box" and "explainable AI"). Facing this landscape of opportunities and challenges, the research group of the *Gallino Laboratory* evaluated the possibility of using and integrating the potential offered by the logic of traditional expert systems with the more modern approaches of machine learning. In the context of curricular education, this approach seems to offer a series of interesting and advantageous potentials. Indeed, these systems allow precise control in the design and use of educational programs, ensuring, from a content perspective, a progressive and detailed encoding of information, and from a process perspective, a disciplined use of logic. This approach offers the benefits of explainability, predictability, and transparency of content, as well as the ability to easily update or reprogram educational materials while maintaining the system's fundamental characteristics. Expert systems follow well-defined rules, based on human knowledge and experience in a given domain. By combining these rules with the machine learning models exploited by the social robot – which are excellent at analyzing and learning from data – a more controllable and accessible knowledge organization system can be obtained. The added value of this approach is its accessibility for teachers and educators, even those with basic computer skills, thanks to the use of intuitive software like Choregraphe. The prototype also highlights the importance of the physical presence of a social robot in terms of inclusivity and the facilitation of learning.

Conversely, the introduction of a robot into the educational context of a class requires special attention. It is crucial to ensure that its presence is well received by the children, especially those with special educational needs. The acceptance and use by teachers must also be well prepared and designed. Indeed, reactions can vary depending on the contexts, individual experiences, and circumstances. Furthermore, it is important to recognize that, to date, social robots are not error-free. Nao may encounter difficulties in image recognition and voice decoding, especially in noisy environments or with children's voices; it may also have problems maintaining balance during some physical activities. For this reason, every program involving a social robot must be carefully monitored and adapted to the specific context of use. In all this, the supervisory role of the teacher or educator remains fundamental, as well as that of the community of researchers and developers, in order to continuously improve such systems and understand their potential and limitations of use.

Authors' contribution statement

G. Antonello, L. Bano, S. Brignone, R. Grimaldi, and S. Palmieri conceived of the idea presented here. S. B. developed Section 1, L. B. developed Section 2, S. P. developed Section 3, G. A. developed Section 4, while R. G. developed the Conclusion. All of the authors discussed the text and revised the final manuscript.

References

Abbasi, N. I., Spitale, M., Anderson, J., Ford, T., Jones, P. B., & Gunes, H. (2022). Can robots help in the evaluation of mental wellbeing in children? An empirical study. In 2022 31st IEEE international conference on robot and human interactive communication (RO-MAN), pp. 1459-1466.

Alnajjar, F., Bartneck, C., Baxter, P., Belpaeme, T., Cappuccio, M., Di Dio, C., Eyssel, F., Handke, J., Mubin, O., Obaid, M., Reich-Stiebert, N., & Reich-Stiebert, N. (2022). Robots in education: An introduction to high-tech social agents, intelligent tutors, and curricular tools. New York and London: Routledge.

Bartneck, C., Belpaeme, T., Eyssel, F., Kanda T., Keijsers, M., & Šabanović, S. (2020). Human-robot interaction: An introduction. Cambridge: Cambridge University Press.

Belpaeme, T. (2002). Human-Robot Interaction. In Cangelosi, A., & Asada, M., (Eds.), Cognitive robotics. Cambridge, MA: MIT Press, pp. 379–393.

Belpaeme, T., & Tanaka, F. (2021). Social robots as educators. In OECD Digital Education Outlook 2021. Pushing the Frontiers with Artificial Intelligence, Blockchain and Robots. Paris: OECD Publishing, pp. 143–158.

Belpaeme, T., Kennedy, J., Ramachandran, A., Scassellati, B., & Tanaka, F. (2018). Social robots for education: A review. In Science robotics, 3(21), pp. 1–9.

Bhaumik, A. (2018). From AI to robotics: mobile, social, and sentient robots. Boca Raton, Florida: CrC Press.

Breazeal, C., Dautenhahn, K., & Kanda, T. (2016). Social robotics. In Siciliano B. Khatib O. (eds.), Springer Handbook of Robotics. Springer, pp. 1935–1971.

Guggemos, J, Seufert, S., Sonderegger, S., & Burkhard, M. (2022). Social robot in education: conceptual overview and a case study of use. In Ifenthaler, D., Isaías, P., Sampson, D. G. (Eds.), Orchestration of Learning Environments in the Digital World, pp 173–195.

Grimaldi, R. (2022) (Ed.). La società dei robot, Milano: Mondadori.

Horner, R. H., Carr, E. G., Halle, J., McGee, G., Odom, S., & Wolery, M. (2005). The use of single-subject research to identify evidence-based practice in special education. In Exceptional Children, 71(2), pp. 165–179.

Korn, O. (Ed.) (2019). Social robots: Technological, societal and ethical aspects of human-robot interaction, Berlin/Heidelberg: Springer Cham.

Lehmann, H., & Rossi, P. G. (2019). Social robots in educational contexts: Developing an application in enactive didactics. In Journal of e-Learning and knowledge Society, 15(2), pp. 27-41.

Lugrin, B., Pelachaud, C., & Traum, D. (2022) (Eds.). The Handbook on Socially Interactive Agents: 20 years of Research on Embodied Conversational Agents, Intelligent Virtual Agents, and Social Robotics, New York: ACM.

Marchetti, A., & Massaro, D. (2023) (Eds.). Robot Sociali e educazione, Milano: Raffaello Cortina.

Meghdari, A., Alemi, M., & Taheri, A. R. (2013). The effects of using humanoid robots for treatment of individuals with autism in Iran. In 6th Neuropsychology Symposium, Tehran, Iran.

Meteyard, L., Cuadrado, S. R., Bahrami, B., & Vigliocco, G. (2012). Coming of age: A review of embodiment and the neuroscience of semantics. In «Cortex», 48(7), pp.788–804.

Mubin, O., Stevens, C. J., Shahid, S., Mahmud, A. A., & Dong, J. J. (2013). A review of the applicability of robots in education. In Technology for Education and Learning, 1(1), pp. 1–7.

Pennisi, P., Tonacci, A., Tartarisco, G., Billeci, L., Ruta, L., Gangemi, S., & Pioggia, G. (2016). Autism and Social Robotics: A Systematic Review. In Autism Research, 9 (2), pp. 165–183.

Ricolfi, L. (1996). Il concetto di modello nelle scienze sociali. In R. Grimaldi (Ed.), Tecniche di ricerca sociale e computer, Torino: Omega, pp. 293-322.

Saleh, M. A., Hanapiah, F. A., & Hashim, H. (2021). Robot applications for autism: a comprehensive review. In Disability and Rehabilitation: Assistive Technology, 16(6), pp. 580-602.

Sarrica, M., Brondi, S., & Fortunati, L. (2020). How many facets does a "social robot" have? A review of scientific and popular definitions online. In Information Technology & People, 33(1), pp. 1–21.

Siciliano, B., & Khatib, O. (Eds.) (2016). Springer Handbook of Robotics, Springer, Cham.

So, W. C., Wong, M. K. Y., Lam, C. K. Y, Lam, W. Y., Chui, A. T. F., Lee, T. L., Ng, H.M., Chan, C. H., & Fok, D. C. W. (2018). Using a social robot to teach gestural recognition and production in children with autism spectrum disorders. In Disability and Rehabilitation: Assistive Technology, 13(6), pp. 527–539.

Tapus, A., Maja, M., & Scassellatti, B. (2007). The Grand Challenges in Socially Assistive Robotics. In Robotics and Automation Magazine, 14(1), pp. 35–42.

Varrasi, S., Di Nuovo, S., Conti, D., & Di Nuovo, A. (2018). A social robot for cognitive assessment. In Companion of the 2018 ACM/IEEE International Conference on Human-Robot Interaction, pp. 269-270.

Woo, H., LeTendre, G. K., Pham-Shouse, T., & Xiong, Y. (2021). The use of social robots in classrooms: A review of field-based studies. In Educational Research Review, 33, pp. 1-11.

Yang, J., & Zhang, B. (2019). Artificial intelligence in intelligent tutoring robots: A systematic review and design guidelines. In Applied Sciences, 9(10), 2078, pp. 1-18.

Youssef, K., Said, S., Alkork, S., & Beyrouthy, T. (2022). A survey on recent advances in social robotics. In Robotics, 11(4), 75, pp.1-28.

Zecca, L., & Datteri, E. (2023) (Eds.). Inclusive science education and robotics: studies and experiences, Milano: Franco Angeli.

Acquiring Language for Human-Machine Interaction: Coding and Computational Thinking from a Logical Point of View

Margherita di Stasio[1], Luca Zanetti[2], Laura Messini[3] and Cristina Coccimiglio[4]

[1] INDIRE, Via Cesare Lombroso 6/15, Firenze, Italy, 0000-0001-6987-4687, m.distasio@indire.it
[2] INDIRE, Via Cesare Lombroso 6/15, Firenze, Italy, 0000-0003-1832-8998, l.zanetti@indire.it
[3] INDIRE, Via Cesare Lombroso 6/15, Firenze, Italy, 0000-0001-8306-8294, l.messini@indire.it
[4] INDIRE, Via Cesare Lombroso 6/15, Firenze, Italy, 0000-0002-4762-8095, c.coccimiglio@indire.it

Abstract

Several types of interactions occur through languages, encompassing both human exchanges and interactions between humans and machines. We are confronted with a reality in which technological interactions are pervasive in everyday life. In this reality, both natural and formal languages play crucial roles, and coding and computational thinking appear to be the paths chosen by educational institutions to address this perspective from a didactic standpoint. This contribution aims to analyze various lists of skills and attitudes discussed in the literature and policy documents. We will present an initial framework developed in exploratory research focused on the philosophy of language and logic as a foundation for coding and valency grammar. Consequently, we will propose a framework that connects logical skills with computational thinking skills.

Keywords

Logical Thinking; Computational Thinking; Coding; Formal Language; Digital Education

1. Introduction

Several types of interactions occur through languages, encompassing both human exchanges and interactions between humans and machines. We confront a reality in which the Internet of Things has rendered technological interactions pervasive in everyday life (van Deursen & Mossberger, 2018), and machines are no longer confined to objects or entities immediately recognizable as robots, computers, or appliances.

Our present context is marked by a unique complexity: over the last century, technological evolution has been profound enough to enable machines to create other machines. Moreover, in the past two decades, machines and technology have become so ubiquitous that expressions such as "infosphere" (Floridi, 2019) or "docusphere" (Ferraris, 2021) have been introduced to capture their extensive influence.

In this reality, both natural and formal languages play crucial roles, and coding and computational thinking appear to be the paths chosen by educational institutions to address this perspective from a didactic standpoint.

This contribution aims to analyse various lists of skills and attitudes discussed in the literature and policy documents. We will present an initial framework developed in exploratory research focused on the philosophy of language and logic as a foundation for coding and valency grammar. Consequently, we will establish a logical skills framework as an analytical tool.

Through this process, we aim to ascertain whether coding and computational thinking can contribute to constructing active participation in the reality described above.

2. Framework and previous experience

2.1. International perspective

Our analysis begins with the *European Reference Framework on the Key Competences for Lifelong Learning*, with a particular focus on the most recent edition (CoE, 2018).

To delve into the role of languages in human exchanges and interactions between humans and machines, we explore the definitions of digital competence and literacy, aligned with the conceptual model of digital citizenship.

On one hand, "Digital competence" is expanded to involve data and media literacy as well as programming included in the broader competence of digital content creation.

On the other hand, "Literacy competence" (formerly "Communication in the mother tongue") has a broad connotation, also considering media and digital materials "across disciplines and contexts" (CoE, 2018).

The need for a broad landscape is underscored in the *Digital Education Action Plan 2021-2027* (CoE, 2020), which asserts that "a solid and scientific understanding of the digital world can build on and complement broader digital skills development. It can also help young people to see the potential and limitations of computing for solving societal challenges". The plan suggests a partnership approach between education, coding, computational thinking, educational robotics, and making – a proposal consistent with recent literature (Chevalier et al., 2020; Scaradozzi et al., 2021).

Regarding the role of computational thinking in schools, one of the most well-known and cited contributions is *Computational thinking: A guide for teachers* (Csizmadia et al., 2015), which views computational thinking as composed of a set of concepts and approaches. The first includes logic, algorithms, decomposition, patterns, abstraction, and evaluation, while the second encompasses tinkering, creating, debugging, persevering, and collaborating. Notably, the authors define computational thinking as "a cognitive or thought process involving logical reasoning" (Csizmadia et al., 2015, p. 6).

In the *Computational Thinking for Teachers* proposed by Commonwealth of Learning, Bhagat and Dasgupta (2021) identified four main components of computational thinking that are: "Decomposition (breaking down a complex problem or structure in to smaller, to manageable elements); Pattern recognition (looking for similarities within problems); Abstraction (focusing only on important information, ignoring irrelevant details); Algorithms (developing a step-by-step solution to the problem, or the rules to solve the problem)" (p. 11).

A recent report, *Reviewing computational thinking in compulsory education* (Bocconi et al., 2022), highlights that several European countries have introduced computer science into curricula, with a particular emphasis on the relationship between "Algorithms" as "problem-solving activities that include the formulation and design of the solution" and "Programming" as the "implementation process" (Bocconi et al., 2022, p. 6). According to the authors, this approach to coding and computational thinking establishes basic skills that are useful for a scientific understanding of the digital reality. Additionally, the report suggests that certain coding activities, particularly setting codified instructions for control robots, can support a different active approach to technology.

Bocconi and colleagues also analyze the relation between computer science competence and computational thinking, confirm the competences earlier identified (Bocconi et al., 2016) and give us an interesting structure composed by:
• abstraction;
• algorithmic thinking, including notion of algorithms and algorithmic problem solving;
• automation, comprehensive of algorithms and programming;
• decomposition;
• debugging;
• generalization, including pattern recognition.

2.2. Italian policy framework

In terms of references to Italian education policies, the picture has become step by step more defined over the past decade. In the documents that define the identity of Italian schools, namely *Indicazioni nazionali* (2012) and *Indicazioni nazionali e nuovi scenari* (2018), we can identify a path that focus a key role of logical and argumentative skills is highlighted.

In *Indicazioni Nazionali* (2012) the reference to coding and related skills is still vague and merely implied (Nulli G., Di Stasio M., 2021). Afterwards, in National Digital School Plan (MIUR, 2015) coding is the focus of a peculiar action (#17). Again in 2015, Law 107, known as "The good school", supports the "development of students' digital skills, with particular regard to computational thinking, the critical and conscious use of social networks and media" (art.1). But only in 2018 *Nuovi Scenari* marks explicit reference to the importance of computational thinking that, as a key skill, must be cultivated as early as elementary school. Computational thinking is defined as education to logical and analytical thinking and is also marked the relations with the educational robotic.

Logic and thinking skills are also crucial in secondary school - across the curriculum and in specific disciplines - and in the tests that universities use to assess enrollment.

2.3. Exploratory research

In 2019 we began an exploratory study focused on the philosophy of language and logic, in relation with History of Philosophy as school subject and with two areas where Indire has an established and peculiar expertise: coding and valency grammar. The aim of the research was to establish the relationship between teachers' logical-philosophical skills and the development of students' thematic and disciplinary skills.

For this purpose, 11 teachers from two *Istituti Omnicomprensivi* (with a total of 10 classes) have been involved in a research and training course.

The research, inspired by Design-Based Research (Design-Based Research Collective, 2003), consisted of four phases in which data are collected whit dedicated methods and tools.

In relation to the national and international policy documents mentioned above, the basic nodes around which the project and the teachers' training was developed were borrowed from the field of philosophy of language and logic: ordinary language and its characteristics; ideal language and its characteristics; the relationship between ordinary and formal language; the relationship between the truth of declarative statements and their form; the role of language in the process of thought; logical connectives.

For valency grammar (will not be covered in this contribution) and coding activities were identified further nodes.

Nodes identified for coding were: definition of coding, use of symbolic languages; algorithms; use of programming language and, on request of the teachers themselves, flowcharts and algorithms.

Table 1
Basic nodes of project and teachers' training

Philosophy Nodes	Coding Nodes
Ordinary language and its characteristics	Definition of coding
The relationship between ordinary and formal languages	The use of symbolic languages
The relationship between the truth of declarative sentences and third form	The use of programming language
The role of language in the process of thought	Algorithms
Logical connectives	Flowcharts and algorithms

At first, teachers studied the nodes of the project through training sessions with the disciplinary experts or individually on texts by Indire. Then they designed a short class activity with the support and supervision of Indire and experts and experimented with the activity in the classroom and eventually redesign. Teachers had also to document the activity so Indire could collect and analyze data.

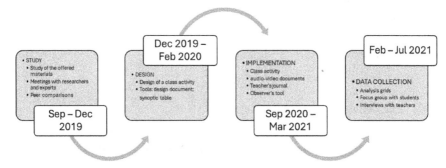

Figure 1: exploratory research timeline

71

Three teachers of primary school chose coding as focus of their short activity: two of them created a 12-hours-project with unplugged activities; one teacher added also a deepening about computer science and coding plugged.

The unplugged activities focused on comprehension, symbols and diagrams and deal with the difficulty of students of following instructions or/and understanding the text of a mathematical problem. Students, divided into groups, created a pixel artwork by writing and following a flowchart.

A class continued the path focusing on algorithms and programming language for a machine (computer), so tried to use software as Blockly, Logo or Code.org

At the conclusion of the exploratory project, it was observed that the themes of Logic and coding that attracted the experimenting teachers the most were:
- the relationship between ordinary and formal language;
- ordinary language and its characteristics;
- formalization;
- logical connectives;
- flowcharts and algorithms.

3. Research questions and methods

We have analysed the competences typically attributed to coding and computational thinking in relation to the fundamental competences for technology awareness and digital citizenship. Additionally, we have summarized the results of our exploratory research.

To actively engage in our reality, young people, in particular, need a broad spectrum of cross-cutting knowledge, attitudes, and skills. This includes classical literacy, digital and data literacy, numeracy, technology, logic, and philosophy.

Our key questions are:
- Can we articulate a basic framework for these skills?
- How can schools support the development of these skills?

We are proposing a framework of logic skills, which we will utilize to analyse the computational thinking skills mentioned earlier and the findings from our exploratory research.

4. Looking for a framework

4.1. A framework for computational thinking

In order to develop a framework, we summarize in a table the skills that emerged from a study of the documents analyzed in §2, along with the purposes for which these skills are highlighted in these documents. From this, we get an overview of a wide array of competences that include data and digital competences, as well as coding/programming and computational thinking.

Table 2
Data and digital competence

Sources	Competences	Purposes
European Reference Framework on the Key Competences for Lifelong Learning (CoE, 2018)	Literacy Competences • Confidence using media (visual, sound/audio) materials • Confidence using Digital Competences • Data literacy • Digital literacy • Programming	Lifelong Learning for successful life in society
Digital Education Action Plan 2021-2027 (CoE, 2020)	Broader digital skills • Coding • Computational thinking • Educational	Scientific understanding of the digital world Understanding of potential and limitations of computing

We then proceed with a specification of the skills related to computational thinking through a comparison of some recent influential frameworks.

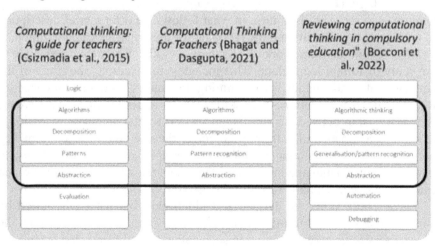

Figure 2: Correspondence between list of computational thinking skills

As it emerges from this comparison, we can consider as basis for our framework the following skills mentioned in all frameworks:
- algorithmic thinking;
- decomposition;
- pattern recognition;
- abstraction.

We will then look for the relationship between these skills and logic. The use of logic as the ground for our framework is consistent with documents on coding and computational thinking, since it is one of the key overarching competences pointed out by Csizmadia and colleagues, and it also plays a fundamental role in the definition of computational thinking assumed in *Indicazioni Nazionali e Nuovi Scenari* (MIUR, 2018).

4.2. A framework for logical skills

We need a framework of logical skills which can function as a ground of skills for the development of skills related to computational thinking and coding. We might have followed different paths to this end. Since the field of logic and the research on logical skills and the teaching of logic have been developed independently from the contemporary interest in computational thinking and coding, the first step of our framework construction was carried out by focusing on logic alone, independently from the connection, if any, that it might turn out to have with computational thinking and coding.

More specifically, in order to identify a framework of skills for logical thinking, three macro categories of skills have been distinguished: skills related to the understanding and use of formal languages (LSs); skills related to formal logic (FLSs); skills related to informal logic (ILSs). Let's briefly explore each set of skills in turn. In what follows we offer a brief description of each category of skills and the rationale for this tripartite categorization.

LSs. Within the framework of LSs there are skills that are not reducible to the study of formal logic alone. This includes, for example, the study of the language of mathematics, programming languages, and the language of formal logic. Among others, these skills include: understanding the difference and relationship between natural language and formal language; knowledge of the characteristics of a formal language; the use and knowledge of key notions such as the notion of function, variable, constant; the ability to translate natural language into formal language; the ability to formulate and use syntactic and semantic rules for the use of formal language.

FLSs. These skills are defined and identified from what is required by university courses in formal logic (and thus from what emerges in the introductory logic textbooks used in such contexts) and from what is required in university admission tests. Even if we confine ourselves to the field of classical logic, there are different understandings and definitions of logic and its goals. (For an introduction, see §1 of Shapiro & Kissel 2022). However, there is a core set of notions and procedures which are shared virtually by all theorists, and which form the basis of any contemporary introductory logic textbook (e.g., Tomassi 1999).

ILSs. The inference rules of classical formal logic are not sufficient to make all the distinctions we want to make in evaluating reasoning in dialogic or dialectical contexts (Goarke 2022). In such contexts, the presence of an interlocutor places constraints on reasoning. Ministerial indications insist a great deal on the importance of learning to reason in a dialogic context. To this end, it is important to identify some specific skills related to informal logic - that is, the kind of logic that studies, among other things, what counts as correct reasoning in dialectical contexts. Such are the skills typically articulated in the context of *critical thinking* skills (for a list see e.g. Ennis 1987).

This tripartite macro-skill subdivision can be used to orient ourselves in the field of competences related to logical thinking. Each macro competence can in turn be subdivided into a plurality of competences - which in turn can be subdivided into hierarchically organised micro competences. Furthermore, competences can be declined according to the order and degree in which their development is to be promoted.

Once we have articulated this tripartite subdivision, our leading question was to know whether there is any relationship between these logical skills and skills related to computational thinking. In order to answer our question, we turned our attention to recent publications on computational thinking. Since interest and research in computational thinking is currently growing exponentially, even if there is no consensus over the exact definition of computational thinking there are some authors who suggest lists of skills and attitudes for computational thinking (e.g., Grover & Pea, 2013; Csizmadia et al., 2015, Bhagapt and Dasgupta 2021; Bocconi et al., 2016 and 2022). These lists do not describe the skills in great details (though see Fallon 2016 for an attempt in this direction).. Given these premises, it is no surprise to find that there is only a little number of contributions that try to define the relationship between computational thinking and logical thinking (e.g., Katerina, 2019; Veenman, 2022). There are some studies that explore the relationship between computational thinking and critical thinking skills and attitudes (Walden et al. 2013, Voskoglou et al. 2012, Kules 2016, Smith, 2021). In order to make a contribution towards bridging this gap, we compared our framework of logical thinking skills with the framework of computational thinking skills that emerge from the comparison of three prominent frameworks (see above, par.4.1).

Finally, another key question in our research was to understand the relationship, if any, between computational thinking skills, logical thinking skills, *and* 'the teachers' node identified in our pilot project. The following table summarizes the correspondences between these three aspects.

Table 3

Relationship between computational thinking skills, logical thinking skills, and 'the teachers' node

Teachers' selected nodes	Computational thinking skills	Types of logical skills
The relationship between ordinary and formal language	Abstraction	All skills, especially LeSs
Ordinary language and its characteristics	Abstraction	All skills, especially LSs
Formalization	All skills, especially Abstraction, Algorithmic Thinking and Pattern recognition	All skills, especially FLSs
Logical connectives	All skills	All skills, especially FLSs and ILSs
Flowcharts and algorithms	All skills, especially Algorithmic Thinking	All skills, especially LSs and FLs

The correspondence between teachers' nodes and skills allows us to appreciate the following points (which are valuable for didactic and educational purposes). Nodes on "ordinary language" and "relationship between ordinary and formal language" can be explored in connection to a variety of logical skills, and this is no chance, since skills related to the understanding of formal language bear a tight connection with these nodes, and since formal and informal logic are also ways to capture, in formal terms, some of the features of ordinary language. The node on "formalization" also plays a prominent role in the table, for formalization is the core of the activity of logic and computational thinking involves formalization. Similar considerations apply to the node "logical connectives", since these are the fundamental blocks for constructing complex sentences, which play a key in almost all the skills related to logical thinking. The last node – "flowcharts and algorithms" – is explicitly connected with computational thinking, but its connection with logical thinking skills is less tight. The analysis shows how this table can be used by teachers who wish to work on logical and computational thinking skills in the light of a prior decision concerning the thematic node to be explored in the classroom.

5. Conclusions

As emphasized in *Theoretical Integration of SKILLS: Towards a New Model of Digital Literacy* (Smahel et al., 2023), coding and programming can be viewed as advanced approaches to software, aligning with earlier works on STEM education.

We have identified a perspective in which coding and computational thinking make sense when integrated into a complex framework of knowledge, attitudes, and skills. This framework requires a robust theoretical foundation, and we posit that Logic can provide such a foundation.

We endeavored to outline a framework that:

• Takes logic as a cultural and cognitive basis for computational thinking skills.

• Equips teachers with practical insights to design learning activities that assist students in gaining awareness of the structures and uses of natural and formal languages.

• Allows teachers to discern the difference between logical and computational skills, particularly demonstrating that logical and analytical thinking cannot be reduced to computational thinking.

Developing computational skills within the context of a broader educational agenda that prioritizes logical thinking skills will not only benefit students and teachers in human exchanges but also in interactions between humans and machines.

Acknowledgements

Margherita Di Stasio wrote sections 2.1, 3 and 4.1; Luca Zanetti developed section 4.2; Laura Messini elaborated section 2.3; Cristina Coccimiglio wrote section 2.2; sections 1 and 5 have been jointly conceived.

References

Bhagat, K. K., & Dasgupta, C. (2021). Computational Thinking for Teachers. Burnaby: Commonwealth of Learning.

Bocconi, S., Chioccariello, A., Dettori, G., Ferrari, A., & Engelhardt, K. (2016). Developing computational thinking in compulsory education-Implications for policy and practice (No. JRC104188). Joint Research Center, Publication Office of European Union.

Bocconi, S., Chioccariello, A., Kampylis, P., Dagienė, V., Wastiau, P., Engelhardt, K., Earp, J., Horvath, M.A., Jasutė, E., Malagoli, C., Masiulionytė-Dagienė, V. and Stupurienė, G., (2022). Reviewing computational thinking in compulsory education: State of play and practices from computing education. Joint Research Center, Publication Office of European Union.

Cansu, F. K., & Cansu, S. K. (2019). An overview of computational thinking. International Journal of Computer Science Education in Schools, 3(1), 17-30.

Chevalier, M., Giang, C., Piatti, A., & Mondada, F. (2020). Fostering computational thinking through educational robotics: A model for creative computational problem solving. International Journal of STEM Education, 7(1), 1-18.

Csizmadia A., Curzon P., Dorling M., Humprheys S., Ng T., Selby C., Woolard J. (2015), Computational thinking. A guide for teachers. Retrieved from: https://community.computingatschool.org.uk/resources/2324/single.

Design-Based Research Collective (2003). Design-based research: An emerging paradigm for educational inquiry. Educational researcher, 32(1), 5-8.

Ennis, R. (1987). A Taxonomy of Critical Thinking Dispositions and Abilities, in Joan Boykoff Baron and Robert J. Sternberg (eds.). Teaching Thinking Skills: Theory and Practice, New York: W. H. Freeman, pp. 9–26.

Falloon, G. (2016). An analysis of young students' thinking when completing basic coding tasks using Scratch Jnr. On the iPad. Journal of Computer Assisted Learning, 32(6), 576-593.

Ferraris M. (2021). L'inconscio artificiale. Bollettino Filosofico, 36, pp. 60-68.

Floridi L. (2019). The logic of information: A theory of philosophy as conceptual design. Oxford: Oxford University Press.

Groarke, L. (2022). Informal Logic, The Stanford Encyclopedia of Philosophy, Edward N. Zalta & Uri Nodelman (eds.), retrieved from https://plato.stanford.edu/archives/win2022/entries/logic-informal/.

Grover, S., & Pea, R. (2013). Computational thinking in K–12: A review of the state of the field. Educational researcher, 42(1), 38-43.

Kules, B. (2016). Computational thinking is critical thinking: Connecting to university discourse, goals, and learning outcomes. Proceedings of the association for information science and technology, 53(1), 1–6.

MIUR (2018), Indicazioni nazionali e nuovi scenari. Retrieved from: https://www.miur.gov.it/documents/20182/0/Indicazioni+nazionali+ e+nuovi+scenari/3234ab16-1f1d-4f34-99a3-319d892a40f2.

MIUR (2012), Indicazioni nazionali per il curricolo della scuola dell'infanzia e del primo ciclo d'istruzione. Annali della Pubblica istruzione. Firenze: Le Monnier.

Nulli G., Di Stasio M. (2017). Coding alla scuola dell'infanzia con docente esperto della scuola primaria. Italian Journal of Educational Technology, 25, 2, pp. 59-65.

Scaradozzi, D., Guasti, L., Di Stasio, M., Miotti, B., Monteriù, A., & Blikstein, P. (2021). Makers at school, educational robotics and innovative learning environments: Research and experiences from FabLearn Italy 2019, in the Italian schools and beyond (p. 376). Springer Nature.

Shapiro, S., Kissel, T. (2022). Classical Logic, The Stanford Encyclopedia of Philosophy (Winter 2022 Edition), Edward N. Zalta & Uri Nodelman (eds.), retrieved from https://plato.stanford.edu/archives/win2022/entries/logic-classical/.

Smahel, D., Mascheroni, G., Livingstone, S., Helsper, E. J., van Deursen, A., Tercova, N., Stoilova, M., Georgiou, M., Machackova, H., & Alho, K. (2023). Theoretical integration of SKILLS: Towards a new model of digital literacy. Zenodo.

Smith, J.M. (2021). Is Computational Thinking Critical Thinking?. In: Wen, Y., et al. Expanding Global Horizons Through Technology Enhanced Language Learning. Lecture Notes in Educational Technology. Springer, Singapore.

Tomassi, P. (1999). *Logic*. New York: Routledge.

van Deursen, A.J.A.M. and Mossberger, K. (2018). Any Thing for Anyone? A New Digital Divide in Internet-of-Things Skills. *Policy & Internet*, 10: 122-140.

Veenman K, Tolboom JLJ and van Beekum O (2022) The relation between computational thinking and logical thinking in the context of robotics education. *Front. Educ.* 7:956901.

Voskoglou, M. G., & Buckley, S. (2012). Problem solving and computational thinking in a learning environment. *arXiv preprint* arXiv:1212.0750.

Walden, J, Doyle, M., Garns H, and Hart, Z. (2013). An informatics perspective on computational thinking. *Proceedings of the 18th ACM conference on Innovation and technology in computer science education* (ITiCSE '13). Association for Computing Machinery, New York, NY, USA, 4–9.

"Stringhe: piccoli numeri in movimento"
A national Italian project combining robotics and psychomotricity against educational poverty

Serena Bignamini[1] and Igor Guida[2]

[1] *Stripes Digitus Lab, International Center of Research and innovation on educational robotics and digital technologies, Via Cristina Belgioioso 171, Milano, Italy, serena.bignamini@pedagogia.it*
[2] *Stripes Digitus Lab, International Center of Research and innovation on educational robotics and digital technologies, Via Cristina Belgioioso 171, Milano, Italy, igor.guida@pedagogia.it*

Abstract

The article is about the project "Stringhe: piccolo numeri in movimento". A national Italian project that combines educational robotics and psychomotricity to face and challenge educational poverty, i.e., a range of social and cultural attitudes that deprive and undermine the possibility of learning, experiencing, developing, and allowing skills, talents and aspirations to flourish freely.

The project aims to support around 2,600 children between the ages of 5 and 11 and their families, who live in at-risk contexts characterized by severe economic and social deprivation in the areas of Milan, Naples and Catania, in Italy.

The project is being developed on two parallel tracks, one related to educational robotics and coding, and the other to psychomotricity activities and introduction to sports.

The key theoretical framework of the STRINGHE project, along with its scientific hypothesis, lies in the idea that the introduction to computational thinking and the learning of concepts related to it are greatly facilitated when anchored in real, physically experienced activities with one's own body and body actions.

The construction of the methodology is also based on the construction of monitoring tools based on seven dimensions of behavior that can be classified in terms of computational thinking.

The article aims to present the innovative methodology that comes out of this project and to show the first results of the research conducted by CNR-ITD (Consiglio Nazionale delle Ricerche-Istituto Tecnologie Didattiche).

Keywords
Educational Poverty, Psychomotricity, Sport, Educational Robotics, Coding, Computational Thinking, Innovative Educational Methodology

1. Project overview

The project "Stringhe – piccoli numeri in movimento" is the first project in Italy to combine digital education and psychomotricity to address educational poverty.

It aims to improve the educational offer of schools and community spaces in the suburban areas through the inclusion and creation of innovative methodologies, in order to help children and families living in conditions of severe economic and social deprivation to face educational poverty.

Facing and challenging "educational poverty", understood as a range of social and cultural attitudes that deprive and undermine the possibility of learning, experiencing, developing and allowing skills, talents and aspirations to flourish freely (Ansa.it, 2022), has meant raising the level in our educational proposals.

The term "poverty" should no longer be exclusively associated with the absence of material goods, but when associated with the word "education", it becomes an issue against which we must equip ourselves with innovative methodologies.

The project aims to support around 2,600 children between the ages of 5 and 11 and their families, who live in at-risk environments in specific cities across Italy: Quarto Oggiaro and Bruzzano in Milan (Northern Italy), Secondigliano and Scampia in Naples and Librino in Catania (Southern Italy).

The four-year project is funded by the grant "Un passo avanti" from Fondazione con il Sud - Con i Bambini Impresa Sociale and involves numerous partners, including:

- Fondazione Mission Bambini, lead institution of the project
- Stripes Coop Sociale, lead partner on digital activities together with Palestra per la Mente
- Fondazione Laureus Italia, lead partner on psychomotricity activities
- CNR Consiglio Nazionale delle Ricerche – Istituto Tecnologie Didattiche Palermo, lead partner on Research
- AVANZI SRL, project evaluator
- Third sector entities: Fondazione Aquilone, Associazione C.E.Lu.S., Associazione Talità Kum Fondazione Maria Anna Sala

- Schools: Circolo Didattico G. Parini, I.C.S. Locatelli-Quasimodo, Ic Cardinale Dusmet, Istituto Comprensivo Cesare Cantù, Istituto Comprensivo Trilussa
- Municipalities (districts) of Milan, Naples, and Catania.

The project was granted funding in the year 2019 and activities were to take place from 2020 to June 2024. Due to the COVID 19 pandemic, the activities as they were planned could not take place. Instead, alternative actions were put in place, which were named "Stringhe in emergenza". This resulted in a postponement of activities and an extension of the end of the project to May 2025.

At the time of writing, we are therefore three-quarters of the way through the project and it has involved 10 pre-schools and primary schools with workshop activities, during school and extra-curricular hours (in schools and in some local meeting places), and other schools in the territory of the Lombardy region.

Families in Milan have also been involved through activities proposing the integrated methodology at "Stringhe Lab", one of the project sites based in our Research Centre Stripes Digitus Lab at MIND Milano Innovation District.

The project is developed on two parallel tracks, one related to educational robotics and coding, and the other to psychomotricity activities and introduction to sports.

The key theoretical framework, and the scientific hypothesis as well, of the Stringhe project lies in the idea that the introduction to computational thinking and the learning of concepts related to it are greatly facilitated by physical activities with one's own body and body actions. As Papert also says, knowledge construction is more effective when learners are engaged in designing meaningful projects (Papert S.,1993).

The entire structure of the Stringhe project rests on the idea that the skills underlying coding and educational robotics, which are generally identified as computational thinking skills, are closely linked or related to psychomotor and social-relational skills. Tracing the traits of the relationship between these skills, introducing them into educational activities that are functional for the purpose of the project, and, above all, providing adequate and reliable evaluation and monitoring tools constitutes one of the key points of the implementation of the Stringhe Project Integrated Methodology.

The methodology is based on the integration of the different disciplines initially focusing on a convergence of objectives (1st and 2nd year of the project) and then on a convergence of themes (3rd and 4th years) to be used as a base for laboratory schedule, also in accordance with the educational themes addressed during the year by the schools.

1.1. Project actions

As already emphasized, the project involves different contexts of children's lives, school and extracurricular, and is structured in several actions that we briefly describe here.

Scuola 4.0

This action involves the activation of workshops in psychomotricity and sports, coding and educational robotics in curricular time within pre-schools and primary schools.

Children from the last year of kindergarten through to grade 5 of primary school (age 10) are involved in this action.

These activities involve the co-presence of the teacher and the experts in the two disciplines, aiming to ensure the definition of Class Educational Projects, consulting and training in the field for teachers. Over the 4 years, there is a gradual shift of activities to teachers with a relative decrease in the presence of the experts in motor and digital education. This transition is guaranteed by the Figure of the Pedagogical Consultant and the School Project Referent and is one of the exit strategies and sustainability of the project.

Scuola aperta 4.0

This action involves the same activities as the previous one, in extra-curricular hours. The afternoon activities are open to more groups formed by children selected in agreement between the Pedagogical Consultant, School Referent for the Project and the school's special needs teacher, thus giving more attention to special needs children. The involvement of children with special educational needs will also allow the Stringhe Technical Table to define structures and teaching methods.

In this way it can assess, from a psycho-pedagogical point of view through the new integrated methodology, the learning and growth of the most fragile children.

Smartgym

These are spaces where motor and technological activities are open to other children, (in addition to those intercepted in schools) and families. Places where innovation is made accessible even to those who have always been precluded.

They represent places of family and childcare within the hardship of the large urban areas of the three cities. Smart Gyms were placed in existing Community Spaces in the target suburbs with the aim of strengthening them.

Teacher training

Teachers of the schools involved with the project can also take advantage of training courses taking place at their educational institutions. Several meetings have been organized for pedagogical consultants and educators of the project aimed at more specialized training and integrated methodology.

Stringhe Lab

Stringhe Lab is the action through which the new "Stringhe combined methodology" - integrating psychomotricity and sports, coding and educational robotics - is defined. CNR (Consiglio Nazionale delle Ricerche - Istituto Tecnologie Didattiche) is coordinating the Technical Table formed by Fondazione Mission Bambini, Stripes Coop sociale, Fondazione Laureus and Associazione Palestra per la Mente and defines the roadmap for collaboration and experimentation among the various disciplines and the launch of the integrated methodology.

Integrated activities combine 1 hour psychomotor activity followed by a 1.5 hour digital education activity on a specific common theme. This is because it has been found from these early years of research that groups that do the motor activity before the digital activity achieve far higher results in terms of attention and skill development than groups that did the opposite.

As we shall see below, the construction of the methodology is also based on the construction of monitoring tools, based on seven areas of behaviour that can be classified in terms of computational thinking:

1. Problem-Solving
2. Algorithmic Thinking
3. Metacognition
4. Body Control and Awareness
5. Acting Autonomously and Responsibly
6. Collaborating and Participating
7. Social-Emotional Skills

In order to make the procedure of collecting and processing evaluation and monitoring data easier, a digital platform has been developed, which is an integral part of the Stringhe monitoring instrumentation.

In this article we aim to show you the innovative methodology of this project to show the first results of the research conducted by CNR-ITD (Consiglio Nazionale delle Ricerche-Istituto Tecnologie Didattiche).

2. Educational poverty and integrated methodology

The project Stringhe has two souls: a scientific and a social one. These two identities are not separated: they are strictly connected. The project's final aim is a construction of a new teaching methodology, in which the bond between the body and educational robotics is the innovative key. But we must not forget why we have been called to experiment and build this new methodology. The only answer is to tackle educational poverty. The validated definition tells us that educational poverty is "the deprivation by children and adolescents of the possibility of learning, experimenting, and freely developing skills, and talents" (Save the Children Italiana Onlus, n.d.).

This deprivation goes beyond the economic dimension. It is a multi-level phenomenon that damages a children's potential development: it limits the possibilities of doing, knowing, understanding, and constructing oneself. It feeds the predetermination of the lack of opportunities.

Where do we encounter this predetermination? Especially in suburban neighbourhoods. In these areas, schools are crucial places in the phase of growth and social inclusion for children. In these areas, sometimes schools are the only places where these children can find opportunities.

In the concrete life of a child, what does educational poverty look like? Educational poverty is a lack of stimuli for building oneself, unawareness of one's own abilities, dropping out of school at an early age, not participating in sport, and having no interests.

And in numbers, what does educational poverty look like? It means that in Italy one in seven teenagers drop out of school (12.7%, the second highest percentage in Europe), that 23.1% of young people aged between 15 and 29 are NEETs (Not in Education, Employment, or Training), that one in 4 children never play sports (3-17 years) (Save the Children, 2021) and 12.3% of children do not have access to a PC or tablet (and the percentage raises to 19% in the South) (Morabito, Mauro & Pratesi, 2021).

2.1. Contexts of the project

The project targets the following peripheral areas of Italy that are characterized by situations of severe deprivation from a social and educational point of view.

- Suburbs of Milan (Quarto Oggiaro - Bruzzano - Niguarda): the three areas are inhabited by about 60,000 inhabitants of which 30% are immigrants. There is in the neighbourhoods a significant presence of popular flats (around 2000) where many families live with different frailties. High immigration rates in schools (50 or 60% in some classes) require the identification of innovative teaching tools that facilitate learning in a multicultural context.
- Suburbs of Naples (Secondigliano - Scampia): they are neighbourhoods known for the Camorra, drug dealing, the high rate of poverty, and juvenile crime. The rate of unemployment as well as the incidence of families with potential economic hardship is the highest in the city (ISTAT, 2017a)
- Suburbs of Catania (Librino): one of the most degraded districts. The population has a poverty rate, widespread crime and

unemployment (43.1%) at record levels. Of the all lacking social services and law enforcement (ISTAT, 2017b). The juvenile crime rate is the highest in the city: most of the minors grow up in a condition of serious marginalization and recruitment in illegal circuits often represents one natural choice, an alternative to the destiny of poverty.

The project Stringhe aims to address the issue of children's educational poverty through the requalification of the educational offer in schools in these areas. Providing the cognitive tools to face the future is the main school's function. In the suburbs involved in the project, the situation is alarming: the school system is in great difficulty, with poor building maintenance, little training, fragmentation of school services, and scarce offer of innovative pedagogical paths. If not supported, these schools can become places for the marginalized, not places of protection and accompaniment to childhood. As Massimiliano Costa, professor of General Pedagogy at the Department of Philosophy and Cultural Heritage at Ca' Foscari University in Venice, argues, the tools to prevent educational poverty are investment in early childhood education services, investing in teacher training and inclusive teaching. These tools can make it possible to address the current educational emergency, as the Stringhe Project and its many objectives want to do: "Rethink the territory as an ecosystem for learning, starting from the design of the physical space at school, widening the field to the surrounding environment, exploiting the digital potential, up to transforming places of deprivation into large areas of learning, resilience and both educational and social change" (Moscatelli, 2023).

The project achieves this by providing access to two fundamental disciplines, *psychomotricity and sports* and *digital education.*

Sports are emphasized because they are infrequently practiced in contexts of high social fragility and it's essential for the development path of every child.

Digital activities are highlighted because nowadays the use of technology is omnipresent, and making it a conscious and not a passive experience can determine a child's future.

Using the body and educational robotics as tools, the goal is to create a methodology that is linked to teaching. A methodology that is also linked to the needs of schools, which is accessible, effective, and replicable.

3. Robots against educational poverty

One of the most frequently asked questions when it comes to educational contexts is "why robotics?".

In fact, robotics is often seen as something very technical and difficult to approach and also not very inclusive. It is a very common conception that makes parents, teachers and educators often reticent about its use with even very young boys and girls. Consider, for example, the target audience of the Stringhe project, which starts at age 5.

In this regard, however, I would like to make a premise with respect to the fact that the approach we take is an educational approach to digital technologies and robotics in particular.

This is the specific goal of Stripes Digitus Lab, our international research centre on educational robotics and digital technologies based at MIND Milano Innovation District.

A place where technology and pedagogy come into connection and where experts and professionals in the digital and psycho-pedagogical fields meet, discuss and work together to develop new ideas, approaches and tools that use robotics as a tool for learning, creativity, collaboration and inclusion.

This approach is also what we bring to the Stringhe project on a daily basis. This specific approach to coding and robotics activities brings a specific contribution to facing educational poverty.

It is about guiding children towards acquiring not so much technical knowledge, but specific life skills.

In fact, robotics activities as well as psychomotricity activities also have the great advantage of being engaging and engagement ensures active and constant participation.

On the other hand, digital activities are a focal point in the current landscape and are very much in line with the technological development we are witnessing.

Today's children have and will increasingly have to deal with technological tools, and it is important that they have access to the hard skills and soft skills necessary to be not just passive users of technology, but also active and critical players in its use and creation.

As emphasized also by TechCamp of Politecnico di Milano "educational robotics is a method in which STEM subjects, i.e. the scientific subjects underlying programming, are learnt in a practical and fun way: one learns to use logic, to solve problems with increasing difficulty, increasing one's ability to form what is called 'computational thinking' in the sector, a goal that is also part of coding" (TechCamp – POLIMI, n.d.).

What we do is to use a tool, pc, tablet, robot that is appealing to the child to go to work on developing the figure of the "computational thinker" as it has been defined by the CNR. Our reference is in fact the idea of "computational thinking" built together during the working tables done with CNR, teachers and pedagogical consultants.

The proposed workshops are therefore part of a broader horizon that is that of the integrated methodology between psychomotricity and digital that aims to work on specific skills related to the 7 dimensions highlighted by the working tables presented to us earlier: problem solving, metacognition, algorithmic thinking, collaboration, body awareness, autonomy and social-emotional skills.

The workshop paths are tailored to different age groups according to a common gradualness that can be summarized as follows:

- *Coding unplugged:* we start with coding activities without technological tools done on paper and with their own bodies by defining a "problem," roles (robot and programmer) and a shared code such as PLAY and STOP cards. In these activities children are asked to solve a problem such as accompanying a partner from point A to point B while avoiding a series of obstacles. We never provide a single solution, but children will be able to find different ways and different solutions and decide which they think is more functional. Within this process, another very important element is also crucial: the debugging process. Alessandro Bogliolo, Professor of Information Processing Systems at the University of Urbino, considers this skill very important in every student's curriculum because it allows them to critically notice programming errors. According to the professor, it allows students to correct errors by identifying themselves with the reasoning process that led to the creation of such a creative process (Raiplay, 2017).

- *Robot discovery*: once the logic behind the computer programming of if-then and the roles of robot and programmer are internalized, children are ready to learn about educational robots. Tools such as Thymio, Ozobot, Matatalab can be used without being connected to a PC and programmed with specific software. They are all tools that already allow them to do activities right out of the box. The encounter with these robots always occurs with an initial moment of autonomous discovery of the tool by the children who, alone or collaborating in groups, apply the scientific method to discover how robots work. The subsequent activities always use these robots as tools to address different topics always placed within a narrative framework. Our activities always give importance and relevance to storytelling and connection with didactic or educational subjects.

- *Coding*: with children around 8 years old and older, we introduce coding and robotics activities that require visual or block programming with programs such as Scratch, always with a view to protagonist in the construction of the training by the children and never providing pre-packaged answers but leaving room for experimentation and error as sources of learning.

Each activity is in turn structured in the following phases:

- welcome and introduction;
- independent discovery or resumption of the tool;
- digital programming;
- reorganisation;
- restitution: reworking in which children deploy their metacognitive skills.

As a brief example, here is one of the activities we propose within the project; it is related to the theme of evolution and is based on the children's book *"Grandmother fish"* by Jonathan Tweet.

This activity is usually proposed to the 3rd grade of primary school (8-year-old boys and girls) and is completed in 2 or 3 meetings.

It is an activity that has a strong storytelling and artistic creation component: it begins with reading the book, which allows us to create the context and bring the children fully into the theme we want to address.

As a second step, the children are offered the autonomous discovery of the Ozobot educational robot.

After discovering how Ozobot's colour-based programming works, the pupils are divided into groups.

Each group has to choose an evolutionary line and represent it graphically and create a track in which to move the robot along this line.

An activity like this, seemingly a very simple one, allows us to work on multiple aspects both related to didactics and to the more relational and emotional side:

- listening skills
- creativity and imagination
- application of the scientific method
- sharing
- creation of knowledge and skills from below
- collaborative learning
- learning by doing
- inclusion
- problem solving
- computer science knowledge

- robotics knowledge

As shown, this type of activity makes it possible to guarantee in socially disadvantaged contexts the quality education appointed by Goal 4 of 2030 Agenda by ensuring the usability of digital educational activities that could not have been offered without the implementation of this project.

4. The contribution of sports to the profile of the computational thinker

Facing "educational poverty" means making some changes in the way we meet and work with children. Thus, we felt it was important to introduce into the debate the concept of sports activities, not only as creators of antibodies against educational poverty, but also as a part of an innovative methodology that would mix it with coding and educational robotics. We are talking about a sport that has special teaching characteristics, because its purpose is not immediate performance, but the enhancement of a path of human maturation that enriches the child and that will eventually lead him or her to both individual and collective performance, but because of something much higher and more important.

Rethinking the sports proposal by placing it within the framework of the profile of the computational thinker has meant above all talking about motor intelligence or rather kinaesthetic intelligence.

"By kinaesthetic intelligence we refer to that set of motor knowledge that we can trace in coordination, balance, skill, speed, strength, elasticity, which, if experienced on the body, if internalized, if made to become part of one's own heritage, will become available as mental qualities applicable to all forms of thought, including the computational one".

Our contribution to the project lies in the conjugation, from a motor and sports perspective, of the seven areas of work that define the computational thinker, which have as their ultimate goal to counter the negative effects of educational poverty. The work in these seven areas is not curative but rather preparatory to the formation of "antibodies" to defend against educational poverty.

The dimensions on which psychomotricity proposals are built are as follows:

- Body awareness
- Creativity, experimentation and sharing
- Individual responsibility and social-emotional skills
- Laterality
- Chromatic, spatial, temporal and numerical codes

- Problem solving
- Collaborating and cooperating

These dimensions derive from a convergence of goals with coding and educational robotics activities that aim to work on the same dimensions thus ensuring daily work on integrating and building the Stringhe methodology.

4.1. Create body-based inner landmarks

The importance of starting with bodywork to counter educational poverty is an idea we have been experimenting with for years. The body, understood as in motion during a sporting act, can be an effective tool for creating a positive self-image. A child's idea of themselves in motion, while acting, contributes to forming the self-image that they will later carry over time. A positive awareness and perception of one's actions will obviously influence the child's self-esteem in a positive way. If this does not happen, if a low self-esteem prevails instead, fuelled by a certain degree of motor illiteracy, the risk will be that we will have a large number of motricity impaired children, with all the attendant repercussions on their psychophysical development.

Another key step in experiencing the body in motion is for each child to confidently experience the discovery of his or her dominant laterality. What until recently was a natural and almost always immediate path using the hands, eye and feet is now in danger of being undermined both by the deprivation of motor experiences that the child is called upon to perform and by the use of both hands with electronic devices. If the process of identifying the dominant limb does not take place with due confidence, the conditions for the child's experience of uncertainty could arise. This, and other negative micro-events, could undermine the child's confidence regarding proprioception of their actions, and a child who is insecure about the responses his or her body makes will be a child at risk in their future development.

The guidance of the educational figure in charge of managing the motor activity becomes essential in this regard. It will be necessary to use a methodology that has as its primary goal the enhancement of children's creativity. To this end, it is essential to convey to the students the idea that there is an adult waiting, ready to listen and, if necessary, accept their creative point of view regarding the resolution of a problem.

The adult then becomes a true facilitator in the development of creativity applied to problem solving. To do this, pilot questions such as "how would you do it?" or "show me and your classmates the most original solution you came up with" will be used in such a way as to transform the lesson into a learning environment that has the children as protagonists.

The referring adult must also support students in solving the problems they propose by parcelling out and serializing the different parts of the problem so that they are less complex.

Another task of the adult is to support the child in retroactive thinking, helping him or her to retrace backward the movements enacted in order to identify the most adaptive actions with respect to the purpose and discard the less effective ones.

Of particular importance is the need to accustom the child to live with serenity the alternation between moments of individual responsibility and moments of sharing in small cooperative groups. This dynamic of continuous transition from the "me" to the "us" becomes fundamental in the acquisition of a mental form that overcomes egocentrism, without, however, flattening the child into a collectivism made up of leaders and gregarious. Everyone is perceived as important the moment they realize that they have something to contribute to problem solving. In the Stringhe class, individual, pair, small group and large group exercises will alternate harmoniously.

One of the purposes of our methodology is to support children as they deal with frustration in the face of failure. It is this feeling of "not being able to do," managed and assimilated in the right way with the help of educators, that will provide the motivational foundation needed to deepen the knowledge necessary for problem solving. The Stringhe methodology makes the unresolved, the error, the self-analysis of nonfunctional choices the motivational fuel that will guide them into the future.

In addition, our approach prefers the creation of shared rules from children's ideas and their interactions at times of sporting activity. The regulatory aspect, if experienced as something imposed and dropped from above or as a set of rules to be studied notionally, risks not fulfilling its task. Therefore, it is necessary to involve the children themselves in the creation of a shared set of rules, starting from their needs, directly experienced during play activities. It will most likely be necessary for children to identify which behaviours are considered improper, because they are dangerous, and which are the criteria for awarding points during the game. Creating shared rules, based on direct experience, makes children much more likely to abide by them during the next phase of play.

All of our interventions aim, as an ultimate goal, to create the conditions such that a range of topics can be addressed initially on a body-based basis and later with interventions supported by educational robotics in an integrated perspective. Of course, it is all related and broken down by different age groups, covering the period from the last year of kindergarten through the entire elementary school cycle.

5. An embodied definition of computational thinking. The Stringhe integrated methodology framework and monitoring system: concepts and tools

As previously explained, the Stringhe project aims to integrate two methodological approaches, psychomotricity and educational robotics, into a single, new methodology. This whole process is obviously preceded, developed and monitored with the help of a careful research project in order to provide a solid theoretical and research foundation.

Once developed, the project involves the promotion and use of the new integrated methodology within curricular teaching activities.

The key theoretical assumption that constitutes the scientific hypothesis of project "Stringhe" lies in the idea that the introduction to computational thinking and the learning of related concepts are facilitated if they are anchored in real activities and physically experienced with one's own body (Città *et al.* 2019).

That is, the entire structure of the project rests on the idea that the skills underlying coding, educational robotics, storytelling, and which are generally identified as computational thinking skills, are closely linked or related to psychomotor and socio-relational skills.

In any learning process, the body plays a crucial role. The relationship between the body, its action and the world (environment) allows the mind to be conceived within a social dimension. Such a relationship connotes intelligence as an adaptive property emerging from the interaction between one or more agents, their actions, and the world.

Drawing out the traits of the relationship between these skills, channeling them into educational activities functional to the project's aims and, above all, providing adequate and reliable evaluation and monitoring tools constitutes one of the key points of the implementation of the "Integrated Stringhe Methodology". Two fundamental questions arise from these assumptions:

A. how to best define the concept of Computational Thinking while keeping faith with the idea of its close relationship with the physical sphere?

B. how to evaluate Computational Thinking in the context of the Stringhe project (a school/educational context)?

The *ante litteram* research showed that having good spatial skills increases the likelihood of performing well in tasks requiring the involvement of Computational Thinking (Città *et al.* 2019).

Instead, with regard to the methodological aspect, it has been identified how Computational Thinking is introduced within the EU educational system (Bocconi *et al.*, 2022). There are three main approaches in the inclusion of this competence within the curricular educational pathway:

- as a cross-curricular theme - basic computer science (CS) concepts are addressed in all subjects, and all teachers share responsibility for developing CT skills.
- as a part of a separate subject - basic CS concepts are taught in a computing-related subject (e.g. Informatics)
- Within other subjects - basic CS concepts are integrated within other curricular subjects (e.g., Maths and Technology)

In the context of the Stringhe project, the path taken to answer these questions had, as its orientation points, the listening, collaboration and support of the key players involved in the monitoring and evaluation processes: teachers, pedagogical consultants, educators and smartgym operators.

Working tables were organised involving a substantial number of primary school teachers involved in the project, the pedagogical consultants, the experts and the smartgym operators. Three different tables were structured on the based on the division of the participants according to their subject areas. Each table mainly pursued four primary objectives: (a) to co-construct a shared definition of computational thinking that took into account the hypotheses set out above and that was appropriate for an educational/didactic context; (b) to identify, in collaboration with and with the guidance of the teachers, elements of Computational Thinking in the different subject areas; (c) to identify observable student behaviours (defined as "performance") that could be categorised as a "manifestation" of computational thinking; (d) to construct a consistent monitoring tool from the indications that emerged from the different subject areas in terms of observable behaviour.

All of this led to the structuring of the Computational Thinking proposal of the Stringhe project.

- Computational thinking is a human skill that is involved in the tangible and structured definition of a process or strategy for solving a problem situation.
- This ability is qualified as a high-level skill because it arises from an interaction between different reasoning processes and gives rise to solution schemes that are deeply rooted in the individual's physical and bodily experience.
- Computational thinking is a complex and emergent skill. It enables each individual to break down a situation/problem into its parts, generalize its solution, represent, describe and translate a resolution process into different languages and/or procedures in full awareness of the paths followed, self-regulate, relate to others in a collaborative

manner, all while managing one's cognitive, emotional and bodily resources to the best of one's ability.

In this regard, the definition of computational thinking provided by the CNR is very interesting, which emphasises how it enables students to tackle problems and thought processes: "Computational thinking provides a powerful framework for studying computer science. Its application, however, goes far beyond of computer science: it is the process of recognising aspects of computation in the world around us, and in applying tools and computer science techniques to understand and reason about systems and processes natural, social and artificial systems and processes. It enables students to tackle problems, break them down into solvable pieces and devise algorithms to solve them [...] The emphasis is clear. It focuses on students carrying out a thinking process, not on producing artefacts or proofs. Computational thinking is the development of thinking skills that contribute to learning and understanding" (Consiglio Nazionale delle Ricerche, 2016).

The valuable work of exchange and comparison of the tables led to the elaboration of an operational definition of computational thinking observable in seven behavioural dimensions:

- *Problem-Solving*: Problem solving refers to a cognitive process aimed at turning a situation into a problem to be solved, for which there are no obvious solution methods. Problem solving, as a complex and multidimensional skill, is in fact composed of a series of skills that are directly linked to the typical stages of problem solving, from knowing how to describe a problem to be tackled to the deployment of the necessary solution strategies.

- *Algorithmic Thinking*: Algorithmic thinking refers to a set of distinct cognitive skills that come into play when one wants to understand, verify, improve, or design a problem-solving procedure expressed through an algorithm, that is, through a finite sequence of instructions that are ordered, unambiguous, and coherent. The dimension of 'algorithmic thinking' is articulated in competences with distinctive and closely related traits such as the ability to identify the basic actions that are needed to implement the solution of a problem (identifying possible basic actions), knowing how to describe the solution of a problem in terms of basic actions (constructing a sequence of actions) and knowing how to create complex actions from simple, basic actions (constructing complex actions). Equally crucial in characterising this dimension are the ability to manipulate constructs such as 'perform x number of times' or 'repeat until it happens...' (iteration), the ability to identify the critical points of a situation and the consequences of certain conditions present (control of sequences of actions), and the ability to represent a solution using a symbolic code, whether formal or non-formal (coding) (Futschek and Moschitz, 2011).

- *Metacognition*: According to Flavell (1979, 1987), the term 'metacognition' denotes the knowledge that the individual develops with respect to his cognitive processes and their functioning, as well as his executive activities that preside over the monitoring of cognitive processes. Flavell divides metacognitive knowledge into three categories: knowledge of person variables, task variables and strategy variables. In particular, from the collaborative construction work on this dimension, it emerges that metacognition involves the acquisition of several skills. First of all, the ability to reconstruct one's reasoning and to identify the essential steps of a process, describing it through different codes. Fundamental skills for the monitoring of cognitive processes are, moreover, the ability to identify the constraints that condition a performance, to be able to identify the internal factors that condition the same and to have the ability to identify errors in the course of the process, knowing how to correct them. The ability to monitor and evaluate one's own performance and knowing how to recognise and evaluate possible alternatives to one's own process complete the range of skills that make metacognition an articulated and complex dimension.

- *Bodily Control and Awareness*: Bodily experience is a multifaceted, mostly unconscious experience that depends on the integration of multisensory information about the body in space. This complex integration occurs between automatic, sensory and bottom-up processes (related to the body schema) with those of a higher order, perceptual and top-down (related to body image) (Gurfinkel and Levick, 1991; Kammers et al., 2006). Pivotal aspects of body control and body awareness are: the possession of basic topological concepts, the possession of basic motor schemes and the ability to use and coordinate different motor schemes. Equally crucial are the ability to perceive space using the point of view of others (visual perspective taking), the ability to manage one's own body and motor schemes according to the rules dictated by the problem situation and the ability to represent with the body and to represent the body.

- *Acting Autonomously and Responsibly*: This dimension calls for the monitoring of autonomy and responsibility skills, which are necessary to be able to deal effectively with a variety of situations, understanding them, dealing with them and reflecting on one's actions to adapt them to unforeseen events and changing conditions. Acting autonomously and responsibly, in particular, means being able to choose and use tools, materials and resources suited to the task (choice of tools) and being able to organise and use time autonomously according to objectives (time management). Equally

97

crucial traits of this dimension are the ability to autonomously select and reorganise prior knowledge in the learning process (selection and organisation of prior knowledge), the possession of an adequate awareness of one's own abilities and the maintenance of commitment even in the event of difficulties (self-efficacy).

- *Collaboration and Participation*: One of the most interesting educational actions to investigate is the degree of co-participation of class members in teaching/learning activities, with the aim of systematically monitoring the aspects that characterise healthy communication and cooperation, and the gradual growth of pupils' communication skills. Key competences in this dimension are knowing how to put oneself in the shoes of others by understanding their needs (empathy), behaving respectfully towards others and diversity (respect for others), intervening and collaborating actively with others in pursuit of a common goal (collaborative group spirit). The ability to handle dissent and criticism and to take into account the other's point of view (constructive management of criticism), recognising and accepting different roles and rules (recognition of roles and rules) and knowing how to express oneself in an assertive manner (assertiveness) complete the range of competences that articulate the dimension of collaborating and participating.

- *Social-Emotional Skills*: The set of items referring to this dimension was collaboratively developed to monitor social-emotional competences, defined as that set of skills, knowledge, behaviours, attitudes and values necessary for each individual to effectively manage their affective, cognitive and social behaviour. More specifically, the key traits of this dimension lie in the ability to perceive and recognise emotions, one's own and those of others, and to know how to express them appropriately in context (emotion recognition), to be able to use emotions to facilitate decision-making (use of emotions), to understand and distinguish different emotions and reactions (emotion-reaction linkage), to link emotions to thought by behaving in a manner consistent with the context (cognition-emotion-behaviour linkage). Equally constitutive, in the context of the 'social-emotional skill' dimension, are the ability to tolerate frustration linked to 'no', i.e. linked to exclusion, failure, non-constructive criticism, boredom, etc. (frustration tolerance), and the ability to cope with group pressure by demonstrating sportsmanship and accepting defeat (social pressure management).

5.1.   Research tools

From this definition of computational thinking and the seven dimensions of computational thinking, several assessment tools have been designed that are to be understood as the organs of a single, consistent monitoring body, the STRINGHE monitoring body.

Three tools were structured to monitor and evaluate the activities implemented by the project: the Tool for the group assessment, the "Stringhe grid" for quantitative assessment and the "Stringhe evaluation grid".

5.1.1. The Stringhe Evaluation Grid

The selected tool used in the monitoring process of project activities is the Stringhe Evaluation Grid. It serves as a device through which to make explicit the criteria for evaluating learners' competencies, both in the cognitive and non-cognitive domains.

The Stringhe evaluation grid is compiled by teachers for each child and identifies six dimensions with related items.

Action/Control: The learner is able to identify basic actions that can be used to implement the solution of a complex problem or situation.

Sequencing: The learner can describe the solution to a problem or complex situation as a sequence of basic actions to be performed.

Compositionality: The learner is able to compose basic actions in order to formalize new complex actions that can be used to solve the complex problem or situation.

Iteration: The learner is able to use iteration constructs (e.g. repeat a number of times, repeat until it happens...) to express sequences of repeated actions in a structured manner.

Conditional: Within the sequence of actions that make up the solution to the problem, the learner identifies the critical points and consequences linked to the presence of certain conditions (if x happens, then y).

Coding: The learner represents the solution through the use of a formal or non-formal symbolic code.

Each dimension is assessed through teachers' completion of a grid, a 7-point Likert scale is completed for each item, where 1 indicates "Low Level of competence" and 7 "high level of competence."

ALGORITHMIC THINKING	Competence	1 - Low level of competence					7 - High level of competence
The learner is able to identify basic actions that can be used to implement the solution of a complex problem or situation	Identifying possible basic actions						
The learner can describe the solution to a problem or complex situation as a sequence of basic actions to be performed	Construction of a sequence of actions						
The learner is able to compose basic actions in order to formalise new complex actions that can be used to solve the complex problem or situation	Construction of complex actions						
The learner is able to use iteration constructs (e.g. repeat a number of times, repeat until it happens...) to express sequences of repeated actions in a structured manner	Iteration						
Within the sequence of actions that make up the solution to the problem, the learner identifies the critical points and consequences linked to the presence of certain conditions (if x happens, then y)	Controlling sequences of actions						
The learner represents the solution through the use of a formal or non-formal symbolic code	Encoding						

Figure 1: Example of an excerpt from the STRINGHE Assessment Grid on the dimension of "Algorithmic Thinking"

Key features of the "Stringhe Assessment Grid" are its intelligibility, explanatory clarity, ability to discriminate reliably and non-arbitrarily, and generalization to different observational samples. The "Stringhe Evaluative Grid" used in the Stringhe project is entirely inspired by these theoretical-methodological assumptions and constructed according to a collaborative co-construction process that involved all stakeholders in the implementation process, from researchers to teachers and support figures such as pedagogical consultants and educators.

5.2.   The Tool for the group assessment

Group observation was carried out by filling out an observation grid, constructed from the "Stringhe Evaluation Grid" tool.

The grid allows monitoring of the 7 identified dimensions that include a total of 43 items.

The observer will be asked to express the percentage of students who possess the level of competence reported in each item (0% to 100%).

The instrument also returns the numerical significance, expressed in the number of students in the class, of the level of proficiency achieved.

This tool is filled out by educators in the context of the School 4.0 and SmartGym actions.

indicate the percentage of students who have achieved the specific competence	COMPETENCE	0%-100%
Algorithmic Thinking		
The learner is able to identify basic actions that can be used to implement the solution of a complex problem or situation	Identifying possible basic actions	
The learner can describe the solution to a problem or complex situation as a sequence of basic actions to be performed	Construction of a sequence of actions	
The learner is able to compose basic actions in order to formalize new complex actions that can be used to solve the complex problem or situation	Construction of complex actions	
The learner is able to use iteration constructs (e.g. repeat a number of times, repeat until it happens...) to express sequences of repeated actions in a structured manner	Iteration	
Within the sequence of actions that make up the solution to the problem, the learner identifies the critical points and consequences linked to the presence of certain conditions (if x happens, then y)	Controlling sequences of actions	
The learner represents the solution through the use of a formal or non-formal symbolic code	Coding	

Figure 2: An excerpt from the group monitoring grid

5.3. The Stringhe grid for quantitative assessment

One of the main outcomes pursued by the project is the improvement of cognitive and non-cognitive skills of participating students.

The "Stringhe Assessment Grid" tool allows monitoring the child's growth process throughout the school year.

In order to track the student's progress throughout the project, a quantitative tool called "Numbers in Motion" has been defined alongside it, which takes the seven dimensions of the rubrics and synthesizes them into a single numerical indicator per dimension.

The single dimension indicator makes it possible to summarize the student's level of competence in absolute terms with respect to the entire growth path. The single indicator is expressed on a numerical scale from 0 to 100. The maximum value (100) refers to the expected goal at the end of the internal Stringhe pathway.

The scale turns out to be divided into seven equi-distributed intervals (0-20, 20-40, 40-60, 60-80 and 80-100) corresponding to the attainable competencies during the five classes of elementary school.

1. Problem-Solving,
2. Algorithmic Thinking
3. Metacognition
4. Body Control and Body Awareness
5. Acting autonomously and responsibly
6. Collaborating and participating
7. Social-emotional skills

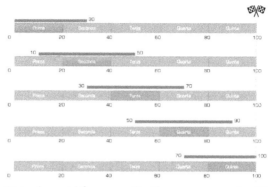

Figure 3: An excerpt from the Stringhe grid for quantitative assessment with individual skill ranges for all primary school classes

5.4. Monitoring Phase

The Stringhe monitoring process involves two different levels of detail:
- group level, which has been used to monitor:
    - the actions where the group involved has a high level of turnover (e.g. Smart-Gym),
    - the actions that will take place at preschools,
    - the actions related to unstructured activities such as events.
- individual student level, which has been used to monitor structured activities, identified mainly in the activities of School 4.0 actions.

In order to verify the effectiveness of the project actions, a control group composed of subjects not taking part in the Stringhe pathway has been defined during the testing process. This has made it possible to determine whether the changes observed and detected in the experimental group are actually attributable to the proposed activities.

Considering the data collected until the end of the 2022-2023 school year, we can say that the monitoring phase has produced the following results in elementary school.

From the project, 1400 students and 249 teachers were involved with a result of 77.16% with respect to quantitative assessment and 77.58% with respect to Evaluation Grid.

Data regarding kindergarten reported the involvement of 171 children and 29 teachers. With a result of 80% of children who possess the level of competencies required at one month after the start of the project activities.

At the time of writing, we are almost at the conclusion of the 2023-2024 school year. The next phases of the research involve collecting data from this school year to go on to modify or consolidate the integrated methodology activities proposed this year and to project into a horizon of thinking that already looks ahead to the project outputs planned for May 2025.

6. Conclusions

This article has attempted to give an overview of the project "Stringhe: piccoli numeri in movimento" by describing the project's objectives, the actions implemented, the two main disciplines involved (psychomotricity and digital education) and the monitoring tools on which the research plan is based.

The project aims to create an integrated methodology between psychomotricity and digital education to face educational poverty, to validate this methodology through the research and the data collected and to create a toolkit that could be spread to all those contexts in which children live in a situation of deprivation.

All the elements described, the direct experience of the operators with the children, and the monitoring data have come together in this year of the project to create the first implementation of the integrated activities and the application of the "Stringhe approach" to schools.

The final year of the project is now beginning, which will see all partners engaged in various actions aimed at concretizing the defined objectives and especially the collection and systematization of data that will come from the monitoring of the year 2023-2024, which will allow them to work on a consolidation of the integrated methodology in view of the last period of implementation and observation and data collection.

Alongside this will be the fundamental importance of teacher training, which by the end of the project will have acquired the necessary skills and knowledge to carry on integrated methodology activities independently in the years to come, thus ensuring the long-term positive impact of the project even after it ends.

Finally, all the experiences and data collected will flow into the creation of a toolkit that will be accessible via the web to a wide target audience of teachers who will be able to take the methodology to their own schools and territories, thus pursuing not only the European directives in education and instruction, but doing innovative teaching and substantially affecting the possibilities and life opportunities of their students.

Acknowledgements

We thank the following colleagues for their contributions to the writing of this article: Irene Villa - Fondazione Mission Bambini, Giovanni Ghidini - Fondazione Laureus; Giuseppe Città, Manuel Gentile, Salvatore Perna - CNR Consiglio Nazionale delle Ricerche Istituto Tecnologie Didattiche Palermo, Andrea Maggioni, Stripes Digitus Lab, Angelo Bianchin, Stripes Digitus Lab.

References

Bocconi, S., Chioccariello, A., Kampylis, P., Dagienė, V., Wastiau, P., Engelhardt, K., Earp, J., Horvath, M.A., Jasutė, E., Malagoli, C., Masiulionytė-Dagienė, V. and Stupurienė, G., (2022) Reviewing Computational Thinking in Compulsory Education, Inamorato Dos Santos, A., Cachia, R., Giannoutsou, N. and Punie, Y. editor(s), Publications Office of the European Union, Luxembourg.

Castoldi, M. (2006). Le rubriche valutative. L'educatore, annata 2006/2007, n. 5.

Città, G., Gentile, M., Allegra, M., Arrigo, M., Conti, D., Ottaviano, S., Sciortino, M. (2019). The effects of mental rotation on computational thinking. *Computers & Education*, *141*, 103613.

Consiglio Nazionale delle Ricerche. (2016). PENSIERO COMPUTAZIONALE una guida per insegnanti. https://tinyurl.com/y3kc8yyw.

Flavell, J. H. (1979). Metacognition and cognitive monitoring: A new area of cognitive–developmental inquiry. *American psychologist*, 34(10), 906.

Flavell, J. H. (1987). Speculations about the nature and development of metacognition. *Metacognition, motivation and understanding*.

Futschek, G., & Moschitz, J. (2011, October). Learning algorithmic thinking with tangible objects eases transition to computer programming. In International conference on informatics in schools: Situation, evolution, and perspectives (pp. 155-164). Springer, Berlin, Heidelberg.

Gurfinkel, V. S., & Levick, Y. S. (1991). Perceptual and automatic aspects of the postural body scheme.

ISTAT. (2017a). *Commissione parlamentare di inchiesta sulle condizioni di sicurezza e sullo stato di degrado delle città*.

ISTAT. (2017b). *Indagine sulle periferie*.

Morabito, C., Mauro, V., & Pratesi, M. (2021). The multidimensional aspects of the educational poverty: a general overview on measures and lack of data in Italy.
From: https://tinyurl.com/5yzm9w87.

Moscatelli, S. (2023). La povertà educativa in Italia: dati, conseguenze e metodi per il contrasto [Blog post]. From:
https://tinyurl.com/27y7fhn7.

McTighe, J., & Ferrara, S. (1996). Performance based Assessment in the classroom: A planning Framework. *A handbook for student performance assessment in an era of restructuring. Alexandria, VA: Association for Supervision and Curriculum Development*, 1-5.

Kammers, M. P., van der Ham, I. J., & Dijkerman, H. C. (2006). Dissociating body representations in healthy individuals: differential effects of a kinaesthetic illusion on perception and action. *Neuropsychologia*, 44(12), 2430-2436.

Papert, S. (1993). *The Children's Machine*.

Povertà educativa in aumento, il futuro sottratto ai giovani - Notizie - Ansa.it [Blog post]. (n.d.). From: https://tinyurl.com/2rcrr442.

Raiplay. (2017). Coding - ToolBox 2 [Video]. From: https://tinyurl.com/bddbmfft.

Save the Children. (2021). GUARANTEEING CHILDREN'S FUTURE How COVID-19, cost-of-living and climate crises affect children in poverty and what governments in Europe need to do. From; https://tinyurl.com/29dhdxk8.

Save the children Italia Onlus. (n.d.). LA LAMPADA DI ALADINO. L'indice di Save the Children per misurare le povertà educative e illuminare il futuro dei bambini in Italia. From https://tinyurl.com/4d5jfu84.

TechCamp - POLIMI. (n.d.). Robotica Educativa Cos'è e perché è importante [Blog post]. From: https://tinyurl.com/24p2vnsx.

MiRo-E at the Scuola Audiofonetica di Brescia – Exploratory analysis for an inclusive approach

Hagen **Lehmann**[1], Federica **Baroni**[2], Laura Sara **Agrati**[3] and Marco **Lazzari**[4]

[1] *Dipartimento di scienze umane e sociali, Università di Bergamo, Piazzale Sant'Agostino, 2, 24124 Bergamo (BG), Italia, hagen.lehmann@unibg.it,*
[2] *Dipartimento di scienze umane e sociali, Università di Bergamo, Piazzale Sant'Agostino, 2, 24124 Bergamo (BG), Italia, federica.baroni@unibg.it,*
[3] *Dipartimento di Psicologia e Scienze dell'Educazione, Università Telematica Pegaso, Centro Direzionale Isola F2, via G. Porzio, 4, 80123 Napoli (NA), Italia, laurasara.agrati@unipegaso.it,*
[4] *Dipartimento di scienze umane e sociali, Università di Bergamo, Piazzale Sant'Agostino, 2, 24124 Bergamo (BG), Italia, marco.lazzari@unibg.it*

Abstract

The work presented in this paper is part of a collaboration between the *Scuola Audiofonetica di Brescia* and the Department of Human and Social Sciences of the University of Bergamo. It describes the introduction of an educational robotic platform, MiRo-E, into a school lessons that include students with hearing disabilities, specific learning disabilities and other special educational needs. The paper will give an introduction to the rationale behind the project and will describe the first exploratory results based on an initial qualitative evaluation.

Keywords

Educational Robotics, Inclusion, MiRo-E, Hearing Disabilities

1. Introduction

In the last two decades more and more educational robotic applications have emerged firmly placing social robots in the realm of effective tools for didactics (Belpaeme et al., 2018). Specific attention in this context has been paid to didactic approaches that promote the inclusion of children with special needs. The proposed solutions range from applications that facilitate the classroom attendance of children with physical impairments (Weibel et al., 2020) to robot-assisted therapy for children with various levels of Autism Spectrum Disorder (Cao et al., 2019).

The work presented here is part of a collaboration between the *Scuola Audiofonetica di Brescia* (hereafter SAB) and the Department of Human and Social Sciences of the University of Bergamo, within the framework of the three-year project "For ALL: Accessibility, Languages, Learning". The topics of this project are inclusive technologies (Lazzari, 2017) and the application of the principles of Universal Design for Learning (Rose et al., 2002) in the production of accessible teaching materials, which also includes interventions based on robotics and coding, for the design and implementation of laboratories for Science, Technology, Engineering, Arts and Mathematics (hereafter STEAM) education (Martinez-Paez et al., 2019) in the first-grade secondary school (*scuola secondaria di primo grado*), and the introduction of coding and educational robotics in activities in nursery and primary school.

The objectives of the research presented in this paper are: (1) an increase in the use of technology in teaching practice through the implementation of robots and principles of computational thinking; (2) the development of digital competences for children in accordance with the European Union competences classified in the current DigCompEdu document (The Digital Competence Framework for Citizens); (3) the inclusion of students with disabilities in educational robotics activities (Lehmann, 2020), both to support logical-mathematical learning and to promote prosocial behavior and communication through the use of social robots as mediators (Besio et al., 2022); and (4) the improvement and extension of the content and experiences of the *Cognitive-Operational Education laboratory* approach *(Laboratorio di Educazione Cognitivistico–Operazionale)*, which is a didactic approach developed specifically for the inclusion of children with hearing disabilities at the SAB.

On a practical level, the aim, in collaboration with the teachers at the school, is to examine what methods and practices can promote the integration of digital technologies and robotics in concrete educational settings (e.g. computer science lessons), to study the effects of the robotic interventions via a structured experimental approach, and the creation of evaluation tools that can help to enhance coding and robotics-focused teaching.

In our work we use the MiRo-E robot, a platform developed by *Consequential Robotics* (www.miro-e.com) specifically for different educational contexts. MiRo-E has different sensors and actuators that enable it to communicate non-verbally and express simple emotional states via movements, and acoustic and visual stimuli. It can interact with its environment to a certain extend autonomously, based on the input of its sensors, or it can be controlled either directly via remote control or together with a dedicated software development environment called MiroCode. This software environment allows teachers and children to write simple programs in a block-based programming language on a work station and then test these programs on the robot. Another important feature of the MiRo-E ecology is MiRoCloud, a dedicated software that allows teachers to easily assign groups of students access to the programming environment and in this to facilitate group work.

As mentioned above, the partner institute for this project is the *SAB*. This school follows an inclusive approach for children with hearing disabilities. It is also the focal point of the development of the *Cognitive-Operational Education laboratory* didactics approach (*laboratorio di Educazione Cognitivistico–Operazionale*) (Scattorelli and Taraschi, 2014), which is based on constructivist principles (Piaget, 1954) and encourages children to learn abstract logical-mathematical concepts via movements and actions. The background of the school gives us the possibility to test the robot not only in inclusion scenarios with children of different age groups but also to experiment with different didactic scenarios. The robot is currently used in the first-grade secondary school (*scuola secondaria di primo grado*), but we plan to use it also in nursery school (*scuola dell'infanzia*) and in primary school (*scuola primaria*) in the next scholastic year. The design of the interaction scenarios is tailored to the curricular requirements of the school and the needs of the children in their specific age groups. Most of the work discussed in this paper is concerned with the development stage of the robots implementation and based on exploratory instructional sessions with the teacher at the school.

We will start by describing the specificities of the *SAB* and the *Cognitive-Operational Education laboratory* approach (*laboratorio di Educazione Cognitivistico–Operazionale*), of which the school is the proponent. In the second part we will describe our methodology, including an overview of the robot and the specific scenarios we are using in class. After this we will present some preliminary results based on an exploratory qualitative approach, in which we used the Metaplan method (Schnelle, 1979) to evaluate the initial attitude of the teachers towards robots. At the end we will discuss future plans and research directions.

2. Context

2.1. The Scuola Audiofonetica di Brescia

The *SAB* is a private institute, managed by the *Fondazione Bresciana per l'Educazione Mons. Giuseppe Cavalleri*, which in the current school year 2023/24 houses 582 pupils aged between 1 and 13 (from nursery to lower secondary school). This number includes 85 children with certified disabilities (62 with prevalent hearing impairments; 23 with other cognitive and learning disabilities). Approximately 26 percent of the deaf students have other associated disabilities and 45 percent come from families of non-Italian origin, which further increases the complexity of the teaching and other associated educational activities. All pupils with disabilities are integrated in ordinary classes, which makes the school a project of Italian national importance in which specialized teaching for deaf students is integrated with general teaching interventions. The school adopts a didactic approach that is driven by experiments in collaboration with the Cedisma center (Centre for Studies on Disability and Marginality) of the Catholic University of the Sacred Heart (Folci, 2020).

The teachers at the *SAB* use personalized teaching-learning processes in classes of 18-24 students, including students with certified disabilities, specific learning disabilities and other special educational needs. To be able to support this complex teaching responsibility, the school has an increased number of teachers and staff, which allows it to make teaching flexible and to personalize interventions. In the current scholastic year 2023-2024, 90 teachers, 8 communication assistants and 24 educators are employed by the school. Working alongside the teachers are a director, three coordinators, a contact person for inclusion and a school psychologist, five speech therapists (for deaf students), an audiologist/speech therapist and an audiometry specialist. The workshops (musical, cognitive-operational, artistic), as well as the flexible teaching methods tailored to small groups, allow for the creation of inclusive teaching that enhances differences and responds to the specific needs of students with disabilities (Scuola Audiofonetica, 2020).

2.2. The Cognitive-Operational Education Laboratory approach

The *Cognitive-Operational Education laboratory* was developed in the *SAB* in the early 1980s, as a result of research based on the observation of the learning strategies of children with hearing impairments, and on the need to propose inclusive teaching activities that strengthen their cognitive abilities in order to overcome the linguistic difficulties linked to more formal approaches in the study of mathematics and logic. The theoretical and practical framework, as well as the first experiments, were established by Pea et al. (Pea, 1987). This early work strongly supported the importance of physical-sensory and concrete experience in learning of topological concepts and in the understanding of spatial and temporal relationships, often difficult to acquire in the early years of nursery and primary school, in particular for deaf students or those with delays in language development.

The activities within the laboratory are offered to all children, deaf and hearing, in a dedicated and specially equipped space, which is large and free of traditional school furniture, so as to allow free movement. Wooden materials (e.g. numbers, letters, circles on the ground, etc.), objects specially arranged for orientation in space activities and reading/formalization of actions, mattresses, musical instruments and other structurally altered materials constitute the setting of the laboratory space.

The children are progressively accompanied on educational paths that develops in three phases:

Phase 1. Experiential phase in which reality is conceptualized via the body;

Phase 2. Symbolic phase in which the concepts become symbols and reality is codified through   multiple representations (visual, auditory, tactile);

Phase 3. Abstraction phase in which the child conceptualizes what s/he has experienced.

According to this approach physical experience is fundamental for the development of personal identity and the cognitive processes that underlie logical-mathematical and spatial-temporal competences (LeBoulch, 1993), furthermore the child needs to be placed in a position to experiment, self-evaluate and correct herself in all phases of the process, experiencing errors as an integral part of the learning process.

Through movement, motor act and action, the student plans sequences of movements aimed at a purpose, deciding if and when to implement them, perceives and structures the body scheme, consolidates space-time orientation, attention, short- and long-term memory; develops logic, imagination and creativity, exercises fine motor skills, executive functions and strengthens social skills.

The approach underlying the *Cognitive-Operational Education laboratory* of the *SAB* has strong theoretical and practical connections with the didactic approach of Seymour Papert (Papert, 1993). Both approaches focus on the prevalence of the physical dimension in the exploration of reality, the use of the body in space, and the active role of the child as protagonist in the learning process. For these same reasons it is easy to find significant correspondences with the most modern practices of unplugged coding and the embodied approach in STEAM learning: in this sense the school experience can be enhanced in the construction of a teaching framework consistent with the proposal of coding and educational robotics activities (Lehmann, 2020).

3. Method

3.1. Robotic Platform MiRo-E

From the beginning of the project, it was clear that a robotic platform was needed that can be adapted to the specificities of the inclusive approach of the *SAB* and to the needs of the different age groups that are involved in the project. Concretely that meant a robot that can express itself with nonverbal signals, such as differently colored lights and intuitively interpretable movements, and that has software which allows for the teaching of basic programming principles.

Considering the particular context situation, characterized by the presence of numerous pupils with disabilities and the notable age difference between the pupils receiving the intervention (from 3 to 13 years), a robot with social robotic characteristics was chosen, one that could be integrated into diverse educational contexts, and that would give both teachers and older children the opportunity to experiment with simple programming environments. Taking these points into consideration the decision was made for MiRo-E, a robotics platform developed by Consequential Robotics in collaboration with the University of Sheffield, UK.

MiRo-E is equipped with several sensors and actuators that allow it to express simple emotional states through movements and acoustic or visual signals. It can be controlled directly via remote control or it can be used in association with a dedicated development environment (MiroCode), which allows children to write simple programs in a block programming language (Blockly and Python) and test them in a 3D simulator before running them on the robot. MiRoCode allows for a variety of different didactic applications ranging from informatics to geometry.

Figure 1: MiRo-E developed by Consequential Robotics

The social characteristics of MiRo-E enable us to use it to encourage collaborative working and to facilitate social communication skills. These elements are an essential part of computational thinking competencies, which are crucial for future generations, so that they can face an increasingly digital world not as passive and unaware consumers of technologies and services, but as active participants in the development of these technologies (DigCompEdu, 2023).

3.2. Initial Evaluation – Attitude towards robots

In order to evaluate on one hand the attitude of the teachers towards digital technologies and robots in education prior to the start of the implementation of MiRo-E, and on the other the knowledge and perception of the children about robots we used the Metaplan® method with 20 teachers and with the students of classes involved in the pilot phase of the experiment (99 in total, attending two classes of primary school and three classes of secondary school).

The Metaplan® method (Schnelle, 1979) is a well-established evaluation approach that facilitates communication and social dynamics within a group of participants discussing a specific topic. It is used to classify and categorize problem dimensions and to weigh the importance of these problem dimensions from the perspective of the topic in question based on the responses of the participants. The strength of this approach is that it uses visualizations of the results of consecutive brainstorming sessions in order reinforce discussions between participants to find categorizations that with the smallest degree of redundancy. During the brainstorming sessions each participant writes down keywords or makes small drawings illustrating their opinion on e.g. post-it notes. These post-it notes are then presented to everyone on a large whiteboard without necessarily making the creator's identity explicit and without creating hierarchies of greater or lesser value between the ideas. From the collection of all the posts, the non-predefined macro-categories that best describe the different ideas that emerged are then identified - with a comparison between participants and under the guidance of the researcher.

4. Results of initial evaluations

4.1. Results from the teacher group

Training meetings were organized for a selected group of teachers (20 teachers in total) from kindergarten, primary and lower secondary schools. The group was formed by bringing together science teachers with previous experience in teaching coding, with teachers that are directly involved in the application of activities within the *cognitive-operational education laboratory* of the *SAB*.

The objective was to discuss the application of educational technologies and social robots in education, with a specific focus on the importance of developing digital competence and a responsible use of digital technologies by students, the opportunity to facilitate inclusive teaching formats and the need to design (and implement) a digital curriculum. During the training, the teachers were introduced to the ministerial lines on computational thinking and digital competences, and to the potential of robotics in reinforcing these competences.

The group, heterogeneous in terms of training, age, professional experience and ability to use IT tools, immediately demonstrated different levels of interest and sharing in the integration of robots in teaching activities, from the enthusiasm for a new line of work, to the concern of lack of competence, up to the refusal to interact with artificial systems. However, even those who showed a resistance to work with robots still took part in the subsequent demonstration session with MiRo-E and participated in the brainstorming, with the Metaplan® method (Schnelle, 1979), aimed at identifying possible scenarios for the introduction of the robot in the classroom of the different age groups.

The core question for the teachers was: "In which concrete scenario do you think this robot could be used in your classroom?".

From the responses of the twenty teachers involved it was possible to identify two groups, with group (a) being more focused on didactic objectives, and group (b) being more oriented towards the concrete classroom settings.

The responses of group (a) can be summarized into three categories:

1. Use of the robot and programming environments to propose teaching activities relating to topological concepts and spatial relationships (Kindergarten and Primary School);
2. Use of the robot and programming environments to facilitate the development of logical-mathematical skills (Primary and Secondary School);
3. Use of the robot and programming environments to encourage the development of social skills.

Regarding the responses of group (b), the idea of working in a small group or with a 1:1 ratio in complex disability situations and in dedicated environments was found to be prevalent; furthermore, some teachers spontaneously highlighted the connections of content and context between the coding and robotics activities and those characterizing the school's cognitive-operational laboratory.

4.2. Results from the student group

After having carried out the training for teachers concerning the basic characteristics of MiRo-E and the MiroCode programming environment, we decided to involve students in pilot classes, to reassure the teachers about the easiness to integrate the robot into the classroom and to offer them opportunities to observe interaction dynamics. These pilot classes involved 99 students from the fifth grade primary and first-grade secondary classes of the school in a demonstration session with the robot and a discussion starting from the stimulus question «What is a robot?», from which reflections emerged on the subject of artificial intelligence, mind-body relationship, social and emotional dimensions, and current and future applications of robots in everyday life.

Also in this case the Metaplan® method was used to facilitate the sharing of ideas within groups homogeneous in age, but heterogeneous in experience, interests, and predisposition to the use of information technologies.

In particular, demonstration and brainstorming sessions were conducted with two primary school groups (45 pupils, including 4 with hearing disabilities and 3 with other disabilities, average age 9 years) and three secondary school groups (54 pupils, including 7 with hearing disabilities and 4 with other disabilities, average age 12 years). Disabilities other than deafness included mild intellectual disability, verbal dyspraxia, Down syndrome, mixed developmental disorder; and children with a Personalized Educational Plan (PDP) for specific learning disabilities or other special educational needs.

From the brainstorming with the students it was possible to identify 7 categories to represent the opinions of the children in response to the stimulus question "What is a robot?":

1. A robot is a form of Artificial Intelligence;
2. A robot is an object with human or animal features;
3. A robot is a useful tool for people with disabilities, who are ill or who need assistance;
4. A robot is an object "to be controlled" (interesting to note the choice of terms attributable to video games and electronic games);
5. A robot is something that will work for us or instead of us in the future;
6. A robot is an object that feels emotions;
7. A robot could be our friend in the future, a companion against loneliness.

From the students' answers it becomes clear that the idea of robots is already a part of the imagination - and in some cases also of the experience - of children of this age group, while Artificial Intelligence is more in their vocabulary than in the reality and knowledge of its meaning or implications.

The social dimension recognized in the robots and the projection of emotions onto technological artefacts points to the necessity to work with the teachers not only from a didactic perspective, but also from an educational perspective to address the underlying ethical issues.

Figure 2: MiRo-E at the Scuola Audiofonetica di Brescia

5. Conclusions

The result of this first exploratory qualitative analysis points the future work of the project in at least two directions:

1) to valorize of the students' potential in the inclusive approach that characterizes the mission of the *SAB* (area 5 of the DigCompEdu framework - 5.1 accessibility and inclusion; 5.2 differentiation and personalization; 5.3 active participation);

2) to support the development of the students' digital skills by helping them to use digital technologies creatively and responsibly (area 6 DigCompEdu).

The challenge lies in defining a setting in which coding activities are perceived by teachers and students as an opportunity to develop computational and critical thinking, where hypotheses, experimentation, observation and verification are the basis of a laboratory approach for discovery, and where programming is understood as a form of expressive-creative language, whether it takes place without the use of computers, through body movements, or whether it takes place on dedicated platforms or with a robot as a mediator. In this sense it is necessary to overcome the logic that sees the humanistic disciplines separated from the natural sciences, or that attributes responsibility for the planning and implementation of STEAM-oriented activities only to teachers of mathematics and natural sciences.

6. Future work

The MiRO-E robots will be implemented on a weekly basis during lessons starting from September 2023 in three different age groups. This process will be closely monitored with the help of different qualitative measures, evaluating the experience and impact of the use of the robot on the attitude and competences of both the teachers and the students.

In the first age group will be children in nursery school (*scuola dell'infanzia*) from 3 to 6 years of age. The approach here will be the enactive robot-assisted didactics approach (Rossi, 2011; Lehmann, 2020) with the main function of the robot being a social mediator and a focal point for shared attention. This will be achieved via different types of group activities and interaction games in which the robot will be controlled remotely by the teacher.

The second group will consist of children in primary school (*scuola primaria*) from 6 to 11 years of age. In this age group, the approach will be the "*laboratorio di educazione operazionale*", using the robot in different physical activities. In these activities, the robot will take on different roles. For the definition of these roles, the classification described by Belpaeme et al. (Belpaeme et al., 2018) will be followed. According to their classification, the most common roles a social robot can assume in educational contexts are the roles of novice, peer and tutor. For the activities at the *SAB*, this means that the robot will solve specific physical tasks, defined by the affordances of the classroom, together with a group of children in such a way that it either pretends to need help, motivates the children to do a task together, or gives instructions to the children. Similar to the previous group also in this group the robot will be controlled remotely.

The third age group will be children in first-grade secondary school (*scuola secondaria di primo grado*) from 11 to 14 years of age. For these children, the

MiRo-E will be used as a tool for STEAM education. The children will use MiRoCode to write small programs during informatics lessons. The topic of these programs will be decided by the teacher, based on the necessities of the school curriculum. After finishing their programs in MiRoCode the students will be able to test them on MiRo-E in the beginning of the following lesson.

References

Appiani, C. (1984). La psicomotricità nell'apprendimento della scrittura, in «HD Giornale italiano di Psicologia e Pedagogia dell'Handicap e delle disabilità di apprendimento», I, 2.

Belpaeme, T., Kennedy, J., Ramachandran, A., Scassellati, B., & Tanaka, F. (2018). Social robots for education: A review. Science robotics, 3(21), eaat5954.

Bers, M.U. (2021). Coding as a Playground, Routledge, New York.

Besio, S., & Bonarini, A. (2022). Robot Play for All. Developing Toys and Games for Disability, Springer, Cham.

Bonaiuti, G., Calvani, A., & Ranieri, M. (2007) Fondamenti di didattica. Teoria e prassi dei dispositivi formativi, Carocci, Roma.

Bonaiuti, G., Calvani, A., Menichetti, L., & Vivanet, G. (2017). Le tecnologie educative, Carocci, Roma.

Canevaro, A., Lippi, G., & Zanelli, P. (1988) Una scuola, uno sfondo, Nicola, Milano.

Cao, H.L., Esteban, P.G., Bartlett, M., Baxter, P., Belpaeme, T., Billing, E., Cai, H., Coeckelbergh, M., Costescu, C., David, D., & De Beir, A. (2019). Robot-enhanced therapy: Development and validation of supervised autonomous robotic system for autism spectrum disorders therapy. IEEE robotics & automation magazine, 26(2), pp.49-58.

Cazzago, P. (1984). Psicomotricità e spazio-tempo. Strutture e ritmi, La Scuola, Brescia.

DigCompEdu (2023) https://joint-research-centre.ec.europa.eu/digcompedu_en

Fedeli, M., & Frison, D. (2018). Methods to facilitate learning processes in different educational contexts, in «Form@re», XVIII, 3.

Folci, I. (2020). L'analisi della qualità del modello organizzativo e pedagogico della Scuola Audiofonetica da parte di CeDisMa, in Scuola Audiofonetica, Sordità e inclusione scolastica, Morcelliana Scholé, Brescia.

Lazzari, M. (2017) Istituzioni di tecnologia didattica, Studium, Roma.

Le Boulch, J. (1993) Educare con il movimento, Armando Editore, Roma.

Lehmann, H. (2020). Social Robots for Enactive Didactics. Franco Angeli.

Martín-Páez, T., Aguilera, D., Perales-Palacios, F.J., & Vílchez-González, J.M. (2019). What are we talking about when we talk about

STEM education? A review of literature, in «Science Education», 103(4).

Papert, S.A. (1993). Mindstorms: Children, computers, and powerful ideas. Basic Books.

Pea, B. (1987) Laboratorio del numero, Emme Edizioni, Torino.

Piaget, J. (1954). The construction of reality in the child. Basic Books.

Rose, D. & Meyer, A. (2002). Teaching every student in the digital age: Universal Design for Learning, ASCD, Alexandria 2002.

Rossi, P.G. (2011). Didattica enattiva. Complessità, teorie dell'azione, professionalità docente: Complessità, teorie dell'azione, professionalità docente. FrancoAngeli.

Scattorelli, F., & Taraschi, M. (2014). Scuola Audiofonetica di Mompiano. Il laboratorio di linguistica operazionale. Imparare a leggere ea scrivere nella scuola primaria.

Schnelle, E. (1979). The Metaplan-method: Communication tools for planning and learning groups, Metaplan, Berlin.

Scuola Audiofonetica (2020). Sordità e inclusione scolastica, Morcelliana Scholé, Brescia.

Weibel, M., Nielsen, M.K.F., Topperzer, M.K., Hammer, N.M., Møller, S.W., Schmiegelow, K., & Bækgaard Larsen, H. (2020). Back to school with telepresence robot technology: A qualitative pilot study about how telepresence robots help school-aged children and adolescents with cancer to remain socially and academically connected with their school classes during treatment. Nurs Open. 2020 Mar 12;7(4):988-997.

Six Hands Robotics

Marco **Binda**[1], Daniele **Brioschi**[2], Ronny **Brusetti**[3] and Emanuela **Scaioli**[4]

[1] *Secondary School Teacher, IC Puercher, Erba (Co), Italy, bindamrc@gmail.com*
[2] *Secondary School Teacher, IC Figino Serenza (Co), Italy, daniele.brioschi@gmail.com*
[3] *Secondary School Teacher, IC Saba,Milano, Italy, ronny.brusetti@gmail.com*
[4] *Retired Secondary School Teacher, Saronno (Va), Italy, emanuela.scaioli@gmail.com*

Abstract

This paper proposes a summary of the "Six-Hands robotics" experience, which was launched in the 2022-2023 school year by four teachers working in different secondary schools, linked by a multi-year teaching and training collaboration.

The teaching of robotics, during curricular hours, and participation in various competitions (WRO, FLL, Amicorobot) have led the teachers involved to reflect on the educational impact they can have on the students involved and the critical issues encountered during robotics competitions (Di Benedetto, 2020).

It has been observed that in such situations competition and motivation to win seem to cloud educational goals. The "Six-Hands Robotics" project identified such an educational need, proposing robotics activities aimed at promoting a different vision, aimed at student participation and involvement, developing positive interdependence and cooperative learning. Unlike other competitions (, during the project events the teams of students competing no longer belong to a single school but form a new "inter-school" team.

Challenges were launched using Lego EV3 kits with mission-based robotics game on home-made fields.

During the events, the teachers carried out observations of the students, focusing on three aspects in particular: emotions and stress management during the competition, in relation to critical and difficult moments; robotics material handling before, during and after the competition; the relational dynamics among peers.

A metacognitive questionnaire was completed by each student after the event. Its function was to help students reflect on what they did during the day and to help teachers evaluate this experience.

The objectives are evaluated with additional questionnaires and focus groups. The real positive impact will arise during the next robotics competitions.

"Six-Hands robotics" project can be replicated and adapted to other contexts by involving a maximum of 3-4 schools. An excessive number of participants would not allow students to fully enjoy the experience and teachers to promote the process monitoring.

Keywords
Educational Robotics, Collaborative Intelligence, Robotics Competition, Team Building, Programming Challenge

1. Introduction

"Six-Hands robotics" experience was launched in the 2022-2023 school year by math and science teachers working in three different secondary schools and linked by a long teaching and training collaboration.

Six-Hands robotics involved 50 pupils, aged 12-13, already participating in educational robotics activities within their school and four teachers.

Educational robotics (Benitti, 2012) is recognized as a valuable means to develop 21st-century skills. It has the potential to promote learning, cognitive and social development, and to increase student engagement in STEM topics. Students taking part in Robotics Challenges build their STEM skills, enhance strong personal skills such as cooperative learning, problem-solving, confidence, quantitative reasoning, computational thinking., creativity, communication (Figure 1) and uncover future STEM careers (Graffin, 2020). The core values are discovery, innovation, inclusion, teamwork and fun.

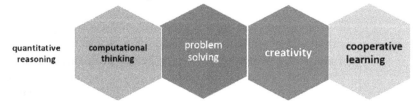

Figure 1: Skills developed in Robotic Challenges

The teaching of robotics, during curricular hours, and the participation (Feng-Kuang Chiang, 2020) in various competitions (WRO, FLL, Amicorobot) have led the teachers to reflect on the educational impact they can have on their students and the critical issues encountered during robotics competitions.

However, it has been observed that in such situations competition and motivation to win seem to cloud educational goals.

Six-hands robotics is a new style of challenge that tries to minimize the competition, keeping the positive aspects of robotics challenges.

Unlike other competitions, during the project events the teams no longer belong to a single school but form a new "inter-school" team. "Six-hands" means that three students from three different schools work in the same group.

Mixing students allows to avoid conflict between different schools; to highlight educational goals and not competitiveness, to boost a task-oriented approach, not a victory-oriented one.

2. Six-hands robotic project

"Six-hands" means that three students from three different schools work in the same group.

The project involved three classes of the three schools with almost 50 students from 12 to 13 years old, already attending robotic courses in their secondary schools. We organized two different Challenges during a whole school day in December and in April.

The sequence was:
1. Definition of goals
2. Planning and setting the competition
3. Definition of group strategy constitution
4. Definition of the challenge themes, challenge fields and missions, with rules and scores
5. Realization of challenge events
6. Monitoring of the project (direct observation, online questionnaire, focus group, etc.)

2.1. Definition of goals

The objectives of the experience mainly concerned four areas, according to:

- Cooperation among peers: promoting the acquisition and consolidation of the social and relational skills necessary to work with unknown peers developing positive interdependence.
- Students' attitudes: developing thoughts and behaviour guided by positive competitiveness in terms of building task-oriented and not victory-oriented personalities.
- Students' participation: making them responsible, able to make decisions independently while face new challenges.
- Involving whole classes: in fact, only small groups of students usually take part in robotics competitions.

2.2. Planning and setting the competition

Teachers' planning was substantially developed online with folder share in Google drive. Each one had different tasks to reach the goals, with a real teamwork.

During the events the venue school produced the physical logistics setup; another school provided part of the already built LEGO EV3 robots and the challenge fields; the last school provided the students' badges, the remaining LEGO EV3 robots that were needed and the competition ruleset datasheet. After each event a questionnaire was prepared for all students. Typical planning of a Challenge is shown in Table 1.

Table 1
Typical planning of a Challenge

Time	Actions
9:50	Launch of the challenge
10:15	Start of the challenge
12:15	First round
12:30	Lunch break
14:30	Second round
15:15	Third round
15:45	Awards ceremony and greetings

2.3. Definition of group strategy constitution and teamwork

The day started with a welcome routine and a simple "*getting-to-know-you game*".

At the end of the game, the badges with the students' names were handed out to facilitate group formation, 17 groups of 3 members each from different schools.

In a confidential file, teachers had drawn a short description of the students to promote the most appropriate pairing. To create the teams, the skills involved in this activity were taken into consideration. Thus, the indicators (Moro, 2011) used to form the groups were:
- interpersonal skills
- resistance to failure
- group management
- programming/construction skills
- Special Educational Needs

Each team was supplied with a robot ready to program the missions of the challenge. Then, each team chose a nickname representing the three students from different schools. During the day, three competition rounds were planned. Between one round and the next, the teams were allowed to modify their robot and their programs to try again any of the missions in the following round. (Figure 2 -3)

During the second event five students were a jury who followed all teams and controlled challenge rules.

Figure 2: Teamwork during Basketball Challenge

Figure 3: Teamwork during Artemis Challenge

2.4.   Challenges and missions

Each challenge, prepared by the teachers and unknown to the students, was launched with an engaging video and then presented and explained to the teams.

The theme of the first event was Basketball, while the second one was Artemis, the new NASA Moon mission.

The first Challenge was launched with a video on a robot playing basketball at the Olympics. (Figure 4) Then the Basketball Challenge was presented and explained to the teams. A second event, Artemis Challenge, used Space as theme with the same didactical approach that is a vision of educational robotics aimed at student participation and involvement, with a view to positive competition. (Figure 5)

Figure 4: Launch of Basketball
Challenge

Figure 5: Launch of Artemis
Challenge

The field (Figure 6) setup was self- built, and each team had one. All teams had three 1.5/2- minute matches to complete as many missions as possible (Figure 7). The rounds were all unconnected and only the highest score mattered. Each team tried to choose the best strategy to obtain the best results. (Figure 8-9)

BASKETBALL CHALLENGE MISSIONS

1. Duffel bag
2. Water bottle
3. Cleaning
4. Ball distributor
5. Shooting at the basket

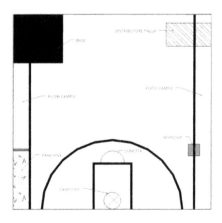

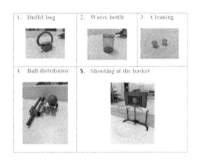

Figure 6: Basketball field project

Figure 7: Robot missions on
Basketball field

Figure 8: Robot in action on Basketball field

Figure 9: A team tries missions on Basketball field

ARTEMIS CHALLENGE MISSIONS (Figure 10-11)

1. Exiting the atmosphere
2. Entering lunar orbit
3. Lunar orbit revolution
4. Return to Earth
5. Moon landing
6. Return of the astronauts to the robot
7. Selfie

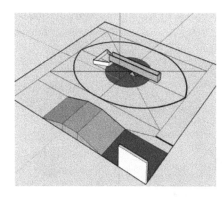

Figure 10: Artemis field project

Figure 11: Robot missions on Artemis field

2.5. Realization of challenge events

The final ranking was made by taking the best round score into account. All the teams tried to reach good results. At first, it was difficult for them to read carefully all the challenge rules, and to choose the best sequence of missions. After some unsuccessful results, a good improvement of their robot and continuous "trial and errors" programming, almost all teams reached good performances. (Figure 12)

Figure 12: Six-hands robotics teams at the end of the Challenge

At the end of the first event, before the greetings, a group photo was printed using a plotter provided by the venue school and all the participants signed it. Some very nice "6-handed robot" gadgets were printed with a laser cutter as a memento of the day and delivered to all participants (Figure 13-14)

At the end of the second event day, "3D-printed Lego Key handlers" were delivered, so each student received a gadget.

Figure 13: Laser cut 6-hand robot gadget

Figure 14: Group photo

2.6. Monitoring

There was an evolution between the two events. During the first event it was difficult for students to read all the challenge rules attentively, and to choose the best sequence of missions. After some unsuccessful results, a good improvement of their robot and continuous "trial and errors" programming, almost all teams reached good performances.

During the second event all teams reached better results in lower time. The rules of the challenge were well explained by a small group of students' juries, directly involved in this role. Students were more comfortable with programming and making a strategy.

A Focus group was organized with five students who were part of the jury during the second competition day. Some of their opinions are written below. (Figure 15)

Figure 15: Voices from Focus Group

A metacognitive questionnaire was completed by each student after each event. Its function was to help students reflect on what they had done during the day and to help teachers evaluate this experience. (Datteri, 2018)

The questionnaire covered the following aspects: planning and construction of the robot, competition, strengths and weaknesses, teamwork, everyone's contribution, organization of the day (Figure 16)

What is your opinion about the results of your group?
(35 answers)

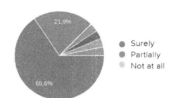

- Awful
- Unsatisfying
- Satisfying
- Well done
- Wonderful

Do you think that you play correctly your role in your group?
(32 answers)

- Surely
- Partially
- Not at all

Figure 16: Data from questionnaires about teamwork

Students from the different schools approached the challenge without rivalry. Most of the groups worked effectively. There were problems with the division of tasks in few cases. They approached the challenges without competitiveness, their work was task oriented (Passalacqua, 2019).

In some cases, the students' jury and the teachers intervened to stimulate the participation of all members and to help solve deadlocks.

3. Conclusion

Six-hands robotics is a pioneering experience inspired this year by four teachers bound by a friendship relationship.

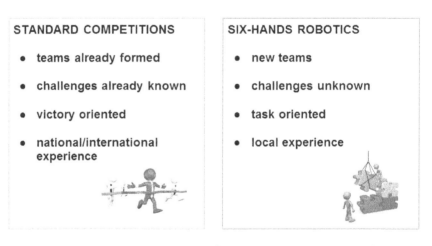

STANDARD COMPETITIONS

- teams already formed
- challenges already known
- victory oriented
- national/international experience

SIX-HANDS ROBOTICS

- new teams
- challenges unknown
- task oriented
- local experience

Figure 17: Comparison between standard robotics competitions and Six-Hands Robotics

The aim was to focus more on educational objectives than on the competition itself. The goals were: cooperation among peer; students' attitudes; participation facing new challenges; involving whole classes.

We think that the four goals concerning these areas were quietly reached. The experience is replicable in schools, and it is adaptable to other contexts, but involving a maximum of 3-4 schools.

An excessive number of participants would not allow students to fully enjoy the experience and teachers to promote the process monitoring. Nowadays there are better and easier chances for schools to join six- hand robotics challenges as an enhancement of the STEM curriculum.

References

Benitti, F.B.V. (2012) Exploring the Educational Potential of Robotics in Schools: A Systematic Review. Comput. Educ. 2012, 58, 978–988.

Bonaiuti, G., Campitiello, L., Di Tore, S., Marras, A., (2020) Educational robotics studies in Italian scientific journals: A systematic review, Frontiers in education.

Datteri E., Zecca L. (2018) Metodi e tecnologie per l'uso educativo e didattico dei robot. Mondo Digitale, 75, 1-6.

Di Benedetto, G. (2020), Presentazione del progetto della rete di scuole lombarde "Amicorobot" e del festival della Robotica Educativa, OPPInformazioni 128, 68-76.

Feng-Kuang Chiang, Yan-qiu Liua,Xiran Fenga,Yaoxian Zhuanga, Yulong Sun (2020), Effects of the world robot Olympiad on the students who participate: a qualitative study, Interactive Learning Environments.

Graffin, M., Shefeld, R., Koul, R. (2022), More than Robots: Reviewing the Impact of the FIRST® LEGO® League Challenge Robotics Competition on School Students' STEM Attitudes, Learning, and Twenty-First Century Skill Development, Journal for STEM Education Research.

Moro, M., Menegatti, E., Sella, F., Perona, M. (2011), Imparare con la robotica educativa, Erikson, Trento.

Part II

Robotics for STEM, language learning and health

Educational robotics for inclusive language learning: theoretical implications and a pilot study with Ozobot

Rita Cersosimo[1] and Valentina Pennazio[2]

[1] *Università di Genova, Laboratorio di Linguaggio e Cognizione, Via Balbi 2, Genova, Italy, 0000-0002-2280-6350, rita.cersosimo@edu.unige.it*
[2] *Università di Genova, Dipartimento di Scienze della Formazione, Corso Podestà 2, Genova, Italy, 0000-0002-3915-1880, valentina.pennazio@unige.it*

Abstract

In this paper, we present a lesson plan that exploits educational robots (i.e., Ozobot) in an English as a Foreign Language (EFL) class. This lesson plan has been tested into a broader project aimed at introducing several innovative technological tools (VR, AI, tablet apps…) to create a more positive, inclusive, and engaging language learning environment (Cersosimo & Pennazio, 2022). 30 children aged 8 to 12 participated in the research, and 8 of them were SEN students diagnosed with dyslexia, learning difficulties, or mild intellectual disabilities. The second lesson of the project was dedicated to educational robotics. We used the Ozobots, small robots that assume the color of the line they are passing over, and whose movements can be programmed with segments of colors. Participants were asked to create in groups an "emotional rollercoaster", i.e., a path on paper in which the Ozobot assumed different emotions through colors and movements (e.g., a robot running fast with a red light on may be interpreted as "angry"). This activity was focused on English vocabulary to express feelings. In the final part of the lesson, each group showed its path to other teams, which had to guess the three emotions expressed by the Ozobot using the newly learned English words. Data collected through two exploratory questionnaires revealed that the activity with Ozobot was one of the most highly appreciated by children if compared to other technological tools, because it created a positive learning atmosphere by mediating the group activity with something physical and easy to use. Participants were also asked to express how they were feeling during the lesson through emoticons, and their answers indicated that positive emotions were associated with the activity. These preliminary results give us the possibility to conclude that using robotics in inclusive language learning education may contribute to creating a positive learning environment, which promotes motivation, self-confidence, and co-

construction of knowledge. Even if actual language learning gains were not the aim of this study, the premises look promising for more deeply investigating inclusive language learning through educational robotics.

Keywords
Robotics, Language Learning, Ozobot, Inclusion, Cooperative Learning, UDL

1. Robotics in language learning

Before integrating robotics into any educational curriculum, it is essential to understand its potential benefits (Johnson, 2003). Robotics has been found to be effective in fostering a wide range of skills, including critical thinking, problem-solving, social interaction, and teamwork abilities (for a comprehensive review, refer to Benitti, 2012), even among individuals with neurodevelopmental disorders (Pivetti et al., 2020). Additionally, robotics has shown promise in enhancing content knowledge across various academic subjects, including science (Altin & Pedaste, 2013; Ayşe & Buyuk, 2021), mathematics (Benitti & Spolaor, 2017; Zhong & Xia, 2020), history (Baxter et al., 2017), and foreign languages (Cersosimo & Pennazio, 2022; Huang & Moore, 2023).

Robotic tools are capable of providing students with an experiential learning opportunity: according to the constructivist and social-interactionist view proposed by Ackermann (2001), educational robotics foster motivation and sociability, while managing to stimulate cognitive, visuoperceptive and motor skills.

In the domain of language acquisition, previous research converges in asserting the crucial role played by active engagement with the tangible, real-world environment in fostering linguistic development (Barsalou, 2008; Hockema & Smith, 2009; Iverson, 2010; Wellsby & Pexman, 2014). Specifically, the enhancement of linguistic proficiency in learners is increased when involving the tangible manipulation of real-world objects and the utilization of bodily movements and gestures (Kersten & Smith, 2002; Mavilidi, Okely, Chandler, Cliff, & Paas, 2015; Rowe & Goldin-Meadow, 2009; Toumpaniari, Loyens, Mavilidi, & Paas, 2015). The positive impact of basic gestures on the language learning process (Glenberg et al., 2011) can be ascribed to two fundamental mechanisms: the enrichment of encoding processes and the increased efficiency in deploying working memory subsystems (Chandler & Tricot, 2015). In accordance with the tenets of embodied cognition, which posit that cognitive processes emerge from purposeful interactions between organisms and their environment (Barsalou, 2008; Glenberg, 1997; Wilson, 2002), the act of responding to information through physical action, rather than passive observation or auditory reception, provides memory with additional cues that are crucial for the representation and retrieval of acquired knowledge. Additionally, the distribution of cognitive load inherent in a learning task across diverse working memory subsystems (visual, auditory, and motor) prevents an excessive burden on any one subsystem (Baddeley, 1992, 2012). This rationale underscores the distinctive potentiality offered by robots in comparison to conventional computer-assisted instructional methods. Robots, exemplified in studies such as Alemi et al. (2014), present novel avenues for pedagogical approaches by facilitating the manipulation of objects and the incorporation of gestures, thereby enhancing language instruction.

In a recent review of research studies that used social robots for language learning, van den Berghe and colleagues (2019) indicate that robots can support grammar learning (Herberg et al., 2015; Kennedy et al., 2016), and the development of reading skills (Gordon et al., 2015; Hong et al., 2016; Hsiao et al., 2015; Hyun et al., 2008), while evidence related to speaking skills is mixed (Hong et al., 2016; In & Han, 2015; Lee et al., 2011; Rosenthal-von der Pütten et al., 2016; Wang et al., 2013). Gains in vocabulary learning have not been pointed out, since a large-scale study by Vogt et al. (2019) suggests no difference between the retention of words taught by a robot tutor and those taught by a tablet application.

One thing the vast majority of research studies seems to agree on is that the presence of robots has positive effects on engagement and attitudes toward what is being learned, no matter the domain or the age group. This is why van den Berghe et al. (2019) conclude that the potential of robots lies mainly in their ability to motivate students. As far as language learning is concerned, robotic tools seem to reduce foreign language anxiety and increase confidence, motivation, and engagement (Alemi, Meghdari, & Ghazisaedy, 2015; Wang et al., 2013; Lee et al., 2011; Eimler et al., 2010; Hsiao et al., 2015; Alemi, Meghdari, & Sadat Haeri, 2017). This can be partly due to the *novelty effect* brought by such robots (van den Berghe et al., 2019), but also to the fact that students appear less anxious about making mistakes in front of a "non-human teacher".

These features are particularly important when it comes to inclusive education because students with Special Educational Needs (SEN) tend to lack self-confidence and engagement in foreign language learning due to the difficulties they usually face (Kontra, 2019; Palladino et al., 2017). However, very few previous studies on robotics in foreign language learning involve this category of people (e.g., Alemi et al., 2015a) and no one mentions the effects of robotics on their engagement in foreign language learning.

2. How to introduce robotics in inclusive designs

The existing array of robotic tools necessitates educators to incorporate them into robust, adaptable, and systematic lesson plans to address students' educational needs (Hockly, 2016). Technologies can serve not only as a functional means for language acquisition but also as components of an inclusive design framework, amplifying benefits in terms of motivation and affectivity. By inclusive design, we refer to an approach suitable for groups of students with diverse abilities, not limited to those with SEN or specific difficulties. Indeed, the initial step in designing any activity, particularly one involving technology, is needs analysis. Depending on the context (school, cooperative, community), this phase should involve team reflection considering the specificities of individuals who will undergo the instructional intervention. Additionally, a careful evaluation of the most suitable tools, the timing of their introduction, the requisite preparation, and especially the methodology employed become essential.

In this context, the principles of Universal Design for Learning (UDL; Savia, 2016) serve as a guiding framework for teachers to adopt a flexible instructional design, allowing for the integration of the digital in its diverse forms. Technology plays a pivotal role in facilitating the implementation of inclusive designs, thus holding significant importance within the UDL paradigm. From the UDL perspective, it is recognized that there is no one-size-fits-all teaching method or tool applicable to all students in every context. Therefore, it is important to provide multiple means of (1) representation (what to learn), involving the presentation of information in various formats, (2) action and expression (how to learn), empowering students to demonstrate their understanding in multiple ways, and (3) engagement (why of learning), involving tasks perceived as meaningful and preferably authentic to encourage students to be actively engaged in seeking answers and solutions (Baroni & Folci, 2022). Historically, the UDL originates from the cultural and architectural movement of Universal Design, which involves anticipating and designing infrastructure, environments, and objects from the outset to be accessible to everyone without any barriers. In the 1990s, the American research group CAST (Center for Applied Special Technology) decided to adapt this approach to the educational context. The undeniable diversity among students makes it much more feasible to design activities that are accessible to all from the beginning, rather than adapting them retrospectively. In this way, the ethical values of equal opportunity and fairness are respected (Rose, 2000), aiming to provide students with equal opportunities for success. In this sense, by "universal", it is not meant that the same method is suitable for everyone, but that everyone has the opportunity to access certain knowledge according to their abilities.

While inclusive design strategies can be implemented within the three principles of Universal Design for Learning (UDL), the instructional framework for integrating technology might be that of a collaborative approach (Bonaiuti, 2014), specifically employing the Cooperative Learning strategy. Cooperative learning encompasses a set of classroom management techniques wherein students collaborate in small groups (Comoglio & Cardoso, 1996). The goal is to enhance learning by leveraging peer cooperation, transcending individualistic or competitive skill concepts to foster social skills—behaviors necessary for constructive interaction with others. According to Johnson and Johnson (1987), the key dimensions characterizing this methodology include social mediation, intentional use of small groups, and cooperative work aimed at maximizing both individual and collective performance. This approach markedly differs from simple group work and, particularly, from traditional teaching centered around the instructor's actions (Morganti & Bocci, 2017). At the core of cooperative learning is the concept of positive interdependence, wherein individuals within the group recognize that their individual success depends on the success of others, and vice versa. This fosters a commitment to effective performance, as individual success is intertwined with collective achievement. For positive interdependence to function, each group member must feel responsible for their contribution to achieving a shared goal. Individual responsibility, therefore, becomes crucial, making each person's contribution indispensable to the group's objectives. This approach helps avoid detrimental phenomena that undermine the inclusive nature of learning environments, such as competition, the presence of free riders (students acting independently), social loafing (delegation and disengagement), and marked differences among class members (Morganti & Bocci, 2017).

In the present study, the UDL methodology was employed to design the integration of robotic tools (as well as other digital tools) in a manner that allowed participants to experience language learning for the first time through a robotic mediator. This mediator was carefully integrated with the instructor's interventions and group interactions beyond the digital context. Consequently, the activities involving robots were organized using cooperative learning, ensuring that the robotic mediator played a facilitative role in fostering collaboration and leveraging diverse skills for the effective execution of the learning journey.

3. Pilot study

3.1. Objectives

Our study exploited educational robots (i.e., Ozobot) in an English as a Foreign Language (EFL) class. This lesson plan has been tested into a broader project aimed at introducing several innovative technological tools such as virtual reality, artificial intelligence, coding and tablet apps, to create a more positive, inclusive, and engaging language learning environment (see Cersosimo & Pennazio, 2022, for a description of the project). For this reason, the theme of emotions was chosen as the guiding thread throughout the entire learning journey. Motivation in English language learning is often lacking in children, particularly those with SEN. One of the primary reasons is attributed to language anxiety that arises during traditional language activities, which often rely on skills in which some of these students may be weaker, such as reading-writing and phonology. Additionally, learning a new language can evoke strong emotions (Palladino et al., 2017) as it involves cognitive, emotional, and motivational effort to access a linguistic and cultural world distant from one's own. Setting the goal of learning the emotions words in English allowed working on them indirectly, making them not only the object of the journey but also a useful tool to achieve the ultimate objective. During the sessions, various moments included reflection on "how one felt" during a particular activity, and doing so in English allowed using the foreign language as a tool, sometimes more indirect and therefore simpler, to communicate one's needs. This choice was also functional to the inclusion of two children from a foster home for minors (see par. 3.2), who often face difficulties expressing feelings, and students with Specific Learning Difficulties (SpLD), who are more prone than others to experience anxiety (Kormos, 2020). The role of emotions in learning has been extensively addressed in the field of inclusive education (Morganti & Bocci, 2017), and neuropsychological research confirms that emotions decisively influence the way learning occurs. Specifically, positive emotions aid in attention and memory, while negative ones impair attention and retention (Clore & Huntsinger, 2007).

3.2. Participants

30 children aged 8 to 12 participated in the research (M=14, F=16; mean age=9.7). 8 of them had a SEN certificate and, more specifically, 6 of them were diagnosed with SpLD, 1 of them had social difficulties (i.e., lived in a foster home for minors), and 1 was in both categories. Participants were recruited through advertisements on Eureka's mailing list and social pages, and by communication sent to local schools. Informed consent was collected from parents or guardians at the inscription, which was free thanks to public funding won by the Cooperativa Sociale Eureka (call "EduCare", Italian Ministry for Family and Equal Opportunities). Cooperativa Sociale Eureka is a learning space in Imperia (Liguria, Italy) that offers activities to support pupils in school subjects and life skills through innovative methodologies and technologies.

3.3. Lesson Plan

Each of the four lessons of the project was designed to present a specific technological tool. We began with a simple software facilitating vocabulary learning coupled with a traditional game, that of charades. Subsequently, the project advanced to encompass robotics, artificial intelligence, and virtual reality. Each weekly meeting, lasting two hours, comprised an initial stimulus phase (warm-up), a central technology-based activity, and a final reflection where children could express their feelings about individual activities using emoticons. Throughout all the sessions, online and offline activities were alternated to ensure a diverse array of stimuli.

The second lesson of the project focused on educational robotics, employing Ozobots, small robots that adopt the color of the line they traverse, and whose movements can be programmed with color segments. Ozobots are considered "minimal social robots" because they are not humanoid nor in animal forms, but their simple form is sufficient to achieve interaction outcomes (Huang & Moore, 2023).

As a warm-up activity, a video from the movie Inside Out was presented. The four segments of the cartoon each represented a feeling of the protagonist, accompanied by a question to elicit vocabulary learned in the previous lesson: "How is she feeling?". Participants, divided into groups of four people according to the cooperative learning principles, were provided with the schema depicted in Figure 1. The fill-in table guided the groups through the activity with the video and the subsequent one, which involved identifying other emotions that could be related to those experienced by the cartoon characters. In line with the UDL principles, students had the option to use the table as a writing and organization aid or to verbally express their thoughts during the brief discussion that followed the vision.

This schema also facilitated the introduction of colors for each emotion, corresponding to the characters in Inside Out and serving a functional role in the next activity.

	Other words that express this feeling
1) How is she feeling? She is _____	
2) How is she feeling? She is _____	
3) How is she feeling? She is _____	
4) How is she feeling? She is _____	

Figure 1: Prompt for the activity

The second part of the session took place with the Ozobots. Each colored segment corresponds to a specific behavior of the robot (direction, speed, special moves). The groups were asked to create an "emotional rollercoaster" for the robot. They had to draw a path in which the Ozobot expressed four different emotions, spreading two functionalities of this device: adopting the color of the drawn lines and following code programming (Figure 2). For instance, a robot running fast with a red light on may be interpreted as "angry". Participants were given the freedom to construct their rollercoaster in their preferred mode, adhering to the principles of UDL. Interestingly, no group opted to create a digital pathway for printing; instead, all groups chose to illustrate their rollercoasters. Within each group, one member assumed the role of the "architect" responsible for designing the path, another acted as the "coder" tasked with communicating the correct code, and one or two members were designated to test the Ozobot and ensure it accurately conveyed the identified emotions.

As a final activity, each group showed its path to other teams, which had to guess the three feelings expressed by the Ozobot.

Figure 2: Children during the activity with Ozobot

3.4. Data collection

At the conclusion of each session, participants were administered a brief questionnaire wherein they used emoticons to assess their emotional experiences during each activity. Emoticons were deliberately chosen to depict various emotions, spanning from sadness to neutrality to happiness. Verbalizing emotions can be challenging, especially for children (Chaplin & Aldao, 2013), and employing emoticons serves as a valuable tool for facilitating easier reflection on these emotions.

3.5. Results

Data analysis involved converting emoticons into numerical values on a scale from 1 to 5, where 1 represented the saddest emoticon and 5 indicated the most joyful. This numerical representation facilitated the calculation of an average score for each activity. Notably, all activities received high scores, ranging from 4 to 5 points (see Figure 3). In particular, after the lesson outlined in this paper, we gathered ratings for both the warm-up activity featuring the Inside Out video and the one involving the Ozobots. These activities received scores of 4.69 and 4.46 points, respectively.

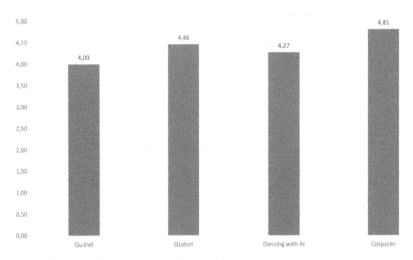

Figure 3: Data from the emoji questionnaire

Informal classroom observation indicated that Ozobot served as an effective facilitator of group activities, outperforming other technologies. According to educators who conducted the lessons, the higher overall score achieved by VR can be attributed primarily to the novelty and engaging features of CoSpaces and the headset visor. Conversely, the positive evaluation of activities involving Ozobot may be attributed primarily to the conducive and friendly environment fostered within the groups. While this observation warrants further systematic analysis in future research studies, it aligns with the notion suggested by several previous research studies that educational robotics can enhance positive experiences when utilized within a cooperative learning framework (Benitti, 2012).

4. Discussion and conclusions

Even if no statistical test was run due to the exploratory nature of the pilot study, we can observe that virtual reality emerges as the most highly appreciated tool, followed by robotics, machine learning, and the more "traditional" flashcard software. While the innovativeness of the tool undoubtedly plays a pivotal role, classroom observations have unveiled another crucial aspect. The ease of use and adaptability to group settings were observed to enhance students' engagement significantly. Activities like Dancing with AI and Quizlet, which required sharing a PC or a tablet among participants, presented challenges in managing cooperative learning roles, potentially contributing to lower evaluations in terms of the emotional aspect of learning. By contrast, the presence of a tangible component alongside the virtual, exemplified by Ozobots, seems to align well with participants' needs and functions as a "mediator" for group activities.

Our study aimed to investigate the potential of various technological tools, including robotics, as motivational and enjoyable elements in language learning, marking the first exploration of its kind. To our knowledge, no prior research has compared the impacts of different technologies in this context. However, aligning with van den Berghe et al.'s (2019) review, we confirmed the positive influence of educational robotics on attitudes toward language learning. While previous studies have focused on broader constructs like motivation and engagement, our specific focus was on emotions, a key aspect of student satisfaction considered integral to engagement (Fredricks et al., 2004). Our findings suggest that robotic tools hold promise in fostering positive emotions even in challenging tasks such as language learning. Notably, our study stands out for utilizing educational robots rather than social humanoid ones, showcasing their efficacy and accessibility.

Yet, it is crucial to recognize that our study was exploratory, and further comprehensive and structured research is warranted. Future investigations should assess the intervention's impact on actual language learning outcomes, incorporating pre and posttests to measure improvements in vocabulary or other skills. Additionally, more rigorous statistical analyses and structured observations of group interactions are needed. For instance, social skills assessment tools like those employed in Mortari et al. (2020), which evaluated conversational skills following a project involving the educational robot Lego Wedo, could offer valuable insights.

The use of simple minimal social robots was a deliberate choice to cater to cooperative needs, but the integration of more sophisticated social robots could potentially enhance language learning in classes with mixed abilities. Given the engaging nature of educational robotics, which sparks curiosity, enables creative learning, boosts student motivation (Alimisis, 2013), and fosters school inclusion (Daniela & Lytras, 2019), we recommend further studies that incorporate robotics into inclusive designs for language learning.

Authors' contribution statement

The article is the result of a close collaboration between the two authors. Specifically, R.C. handled methodology, data collection, data analysis, writing (part 1 to 4). V.P. handled methodology, study supervision, manuscript review and editing.

References

Ackermann, E. (2001). Constructivisme et constructionnisme: quelle difference. In Proceedings of the Conference "Constructivismes: usages et perspectives en education". 1, 85-97.

Alemi, M., Meghdari, A., & Ghazisaedy, M. (2014). Employing humanoid robots for teaching English language in Iranian junior high-schools. International Journal of Humanoid Robotics, 11, 1450022-1–1450022-25.

Alemi, M., Meghdari, A., & Ghazisaedy, M. (2015). The impact of social robotics on L2 learners' anxiety and attitude in English vocabulary acquisition. International Journal of Social Robotics, 7, 523–535.

Alemi, M., Meghdari, A., Mahboub Basiri, N., & Taheri, A. (2015a). The effect of applying humanoid robots as teacher assistants to help Iranian autistic pupils learn English as a foreign language. In Proceedings of the International Conference on Social Robotics 2015 (pp. 1–10). Basel, Switzerland: Springer International.

Alemi, M., Meghdari, A., & Sadat Haeri, N. (2017). Young EFL learners' attitude towards RALL: An observational study focusing on motivation, anxiety, and interaction. In Proceedings of the International Conference on Social Robotics 2017 (pp. 252–261). Basel, Switzerland: Springer International.

Alimisis, D. (2013). Educational robotics: Open questions and new challenges. Themes in Science and Technology Education, 6(1), 63-71.

Altin, H., & Pedaste, M. (2013). Learning approaches to applying robotics in science education. Journal of baltic science education, 12(3), 365.

Ayşe, K. O. Ç., & Buyuk, U. (2021). Effect of robotics technology in science education on scientific creativity and attitude development. Journal of Turkish Science Education, 18(1), 54-72.

Baddeley, A. (1992). Working memory. Science, 255(5044), 556–559.

Baddeley, A. (2012). Working memory: theories, models, and controversies. Annual Review of Psychology, 63, 1–29.

Barsalou, L. W. (2008). Grounded cognition. Annual Review of Psychology, 59, 617645.

Baroni F., & Folci I. (2022). Managing inclusion between Differentiation and Universal Design for Learning: Approaches, Opportunities and Perspectives. Italian Journal of Special Education for Inclusion, X, 2, 61-70.

Baxter, P., Ashurst, E., Read, R., Kennedy, J., & Belpaeme, T. (2017). Robot education peers in a situated primary school study: Personalisation promotes child learning. PloS one, 12(5), e0178126.

Benitti, F. B. V. (2012). Exploring the educational potential of robotics in schools: A systematic review. Computers & Education, 58(3), 978–988.

Bonaiuti, G. (2014). Le strategie didattiche. Roma: Carocci Editore.

Cersosimo, R., Pennazio, V. (2022). "L'inglese tra tecnologie ed emozioni": un percorso inclusivo di avvicinamento alla lingua inglese con elementi di robotica, intelligenza artificiale e realtà virtuale. Lend – Lingua e Nuova Didattica. 4/2022, pp. 40-51

Chandler, P., & Tricot, A. (2015). Mind your body: The essential role of body movements in children's learning. Educational Psychology Review, 27, 365-370.

Chaplin, T. M., & Aldao, A. (2013). Gender differences in emotion expression in children: A meta-analytic review. Psychological Bulletin, 139, 735–765.

Clore, G. L., & Huntsinger, J. R. (2007). How emotions inform judgment and regulate thought. Trends in cognitive sciences, 11(9), 393-399.

Comoglio, M., Cardoso, M. A. (1996). Insegnare e apprendere in gruppo, Roma: LAS.

Daniela, L., & Lytras, M. D. (2019). Educational robotics for inclusive education. Technology, Knowledge and Learning, 24, 219-225.

Eimler, S., von der Pütten, A., & Schächtle, U. (2010). Following the white rabbit - A robot rabbit as vocabulary trainer for beginners of English. In G. Leitner, M. Hitz, & A. Holzinger (Eds.), Proceedings of the 6th Symposium of the Workgroup HumanComputer Interaction and Usability Engineering (pp. 322–339). Berlin, Germany: Springer.

Fredricks, J. A., Blumenfeld, P. C., & Paris, A. H. (2004). School engagement: Potential of the concept, state of the evidence. Review of educational research, 74(1), 59-109.

Glenberg, A. M. (1997). What memory is for: creating meaning in the service of action. Behavioral and Brain Sciences, 20(01), 41–50.

Glenberg, A. M., Goldberg, A. B., & Zhu, X. (2011). Improving early reading comprehension using embodied CAI. Instructional Science, 39, 27–39.

Gordon, G., Breazeal, C., & Engel, S. (2015). Can children catch curiosity from a social robot? In Proceedings of the Tenth Annual ACM/IEEE International Conference on Human-Robot Interaction (pp. 91–98). New York, NY: ACM.

Herberg, J. S., Feller, S., Yengin, I., & Saerbeck, M. (2015). Robot watchfulness hinders learning performance. In Proceedings of the 24th IEEE International Symposium on Robot and Human Interactive Communication (RO-MAN) (pp. 153160). Los Alamitos, CA: IEEE.

Hockema, S. A., & Smith, L. B. (2009). Learning your language, outside-in and insideout. Linguistics, 47, 453–479.

Hockly, N. (2016). Special educational needs and technology in language learning. ELT Journal, 70(3), 332-338.

Hong, Z.-W., Huang, Y.-M., Hsu, M., & Shen, W.-W. (2016). Authoring robot-assisted instructional materials for improving learning performance and motivation in EFL classrooms. Educational Technology & Society, 19, 337–349.

Hsiao, H.-S., Chang, C.-S., Lin, C.-Y., & Hsu, H.-L. (2015). "iRobiQ": The influence of bidirectional interaction on kindergarteners' reading motivation, literacy, and behavior. Interactive Learning Environments, 23, 269–292.

Huang, G., & Moore, R. K. (2023). Using social robots for language learning: Are we there yet? Journal of China Computer-Assisted Language Learning, 3(1), 208–230.

Hyun, E., Kim, S., Jang, S., & Park, S. (2008). Comparative study of effects of language instruction program using intelligence robot and multimedia on linguistic ability of young children. In Proceedings of the 17th IEEE International Symposium on Robot and Human Interactive Communication (pp. 187–192). Los Alamitos, CA: IEEE.

In, J.-Y., & Han, J.-H. (2015). The acoustic-phonetic change of English learners in robot assisted learning. In Proceedings of the Tenth Annual ACM/IEEE International Conference on Human-Robot Interaction (pp. 39–40). New York, NY: ACM.

Iverson, J. M. (2010). Developing language in a developing body: The relationship between motor development and language development. Journal of Child Language, 37, 229–261.

Johnson, D. W., & Johnson, R. T. (1987). Learning together and alone: Cooperative, competitive, and individualistic learning. Prentice-Hall, Inc.

Johnson, J. (2003). Children, robotics and education. In Proceedings of 7th international symposium on artificial life and robotics (Vol. 7, pp. 16–21), Oita, Japan.

Kennedy, J., Baxter, P., Senft, E., & Belpaeme, T. (2016). Social robot tutoring for child second language learning. In ACM/IEEE International Conference on Human-Robot Interaction, 2016-April, 231-238.

Kersten, A. W., & Smith, L. B. (2002). Attention to novel objects during verb learning. Child Development, 73, 93–109.

Kontra, E. H. (2019). The L2 Motivation of Learners with Special Educational Needs. The Palgrave Handbook of Motivation for Language Learning, 495–513.

Kormos, J. (2020). Specific learning difficulties in second language learning and teaching. Language Teaching, 53(2), 129–143.

Lee, S., Noh, H., Lee, J., Lee, K., Lee, G. G., Sagong, S., & Kim, M. (2011). On the effectiveness of Robot-Assisted Language Learning. In ReCALL, 23(1), 25-58.

Mavilidi, M.-F., Okely, A. D., Chandler, P., Cliff, D. P., & Paas, F. (2015). Effects of integrated physical exercises and gestures on preschool children's foreign language vocabulary learning. Educational Psychology Review, 27, 413–426.

Morganti, A., & Bocci, F. (2017). Didattica Inclusiva nella Scuola Primaria. Firenze: Giunti EDU.

Mortari, L., Roberta, Silva, & Zanotti, A. (2020). Quando il Service Learning pone la ricerca educativa a servizio della formazione docente e dell'innovazione didattica: Il "caso" Resolving Robots. RicercAzione, 1, 83.

Palladino P., Botto M., Bellagamba I., Ferrari M., Cornoldi C. (2017). English is fun! : programma per la valutazione degli atteggiamenti e delle abilità nell'apprendimento della lingua inglese. Trento: Erickson.

Pivetti, M., Di Battista, S., Agatolio, F., Simaku, B., Moro, M., & Menegatti, E. (2020). Educational Robotics for children with neurodevelopmental disorders: A systematic review. Heliyon, 6(10), e05160.

Rose D. (2000). Universal design for learning, Journal of Special Education Technology, pp. 45-49.

Rosenthal-von der Pütten, A. M., Straßmann, C., & Krämer, N. C. (2016). Robots or agents—Neither helps you more or less during second language acquisition. Experimental study on the effects of embodiment and type of speech output on evaluation and alignment. In Proceedings of the International Conference on Intelligent Virtual Agents (pp. 256–268). New York, NY: Springer.

Rowe, M. L., & Goldin-Meadow, S. (2009). Differences in early gesture explain SES disparities in child vocabulary size at school entry. Science, 323, 951–954.

Savia G. (eds.) (2016). Universal Design for Learning. Trento: Erickson.

Toumpaniari, K., Loyens, S., Mavilidi, M.-F., & Paas, F. (2015). Preschool children's foreign language vocabulary learning by embodying words through physical activity and gesturing. Educational Psychology Review, 27, 445–456.

van den Berghe, R., Verhagen, J., Oudgenoeg-Paz, O., van der Ven, S., & Leseman, P. (2019). Social Robots for Language Learning: A Review. Review of Educational Research, 89(2), 259–295.

Vogt, P., van den Berghe, R., de Haas, M., Hoffman, L., Kanero, J., Mamus, E., Montanier, J.-M., Oranc, C., Oudgenoeg-Paz, O., Garcia, D. H., Papadopoulos, F., Schodde, T., Verhagen, J., Wallbridgell, C. D., Willemsen, B., de Wit, J., Belpaeme, T., Goksun, T., Kopp, S., ...

Pandey, A. K. (2019). Second Language Tutoring Using Social Robots: A Large-Scale Study. 2019 14th ACM/IEEE International Conference on Human-Robot Interaction (HRI), 497–505.

Wang, Y. H., Young, S. S.-C., & Jang, J.-S. R. (2013). using tangible companions for enhancing learning English conversation. Journal of Educational Technology & Society, 16, 296–309.

Wilson, M. (2002). Six views of embodied cognition. Psychonomic Bulletin & Review, 9(4), 625–636.

Zhong, B., & Xia, L. (2020). A systematic review on exploring the potential of educational robotics in mathematics education. International Journal of Science and Mathematics Education, 18(1), 79-101.

"At School with Pepper": A pilot Study on the Role of Pepper in Learning English as a Second Language

Taziana **Giusti**[1], Alberto **Bacchin**[2], Gloria **Beraldo**[3], Monica **Pivetti**[4] and Emanuele **Menegatti**[5]

[1] The 3° Istituto Comprensivo of Padova, Via Filippo Lippi, 11, 35134 Padova, 0000-0003-2162-6777, taziana.giusti@icbriosco.edu.it

[2] Department of Information Engineering, University of Padova, Via G. Gradenigo, 6/b, 35131, Padova, Italy, 0000-0002-2945-8758, bacchinalb@dei.unipd.it

[3] Department of Information Engineering, University of Padova, Via G. Gradenigo, 6/b, 35131, Padova, Italy, and Institute of Cognitive Sciences and Technologies, National Research Council, via Giandomenico Romagnosi n. 18/A, 00196, Rome, Italy, 0000-0001-8937-9739, gloria.beraldo@unipd.it

[4] Department of Human and Social Sciences, University of Bergamo, Piazzale S. Agostino, 2, 24129 Bergamo, Italy, 0000-0002-8378-2911, monica.pivetti@unibg.it

[5] Department of Information Engineering, University of Padova, Via G. Gradenigo, 6/b, 35131, Padova, Italy, 0000-0001-5794-9979, bacchinalb@dei.unipd.it

Abstract

This paper investigated whether primary school pupils involved in a robot-assisted language learning (RALL) activities with Pepper would outperform their peers in traditional classes in learning English family-related words. Educational robotics can increase students' learning motivation as well as improve students' learning outcomes. Previous meta-analysis showed that RALL outperforms non-RALL conditions, that is RALL has a positive treatment effects on language learning when compared to traditional classroom.

A quasi-experimental research designs was used to test the hypothesis. Two third grade classes were involved: the experimental group of 23 pupils and the control group of 19 pupils. The experimental class interacted with Pepper for a two-hour class during school time in Spring 2022. Pupils from both classes took the same test before and after the intervention. The written test covered the topic of names of family relations (e.g. uncle, grandmother etc). The final score ranged from 0 to 5 in the pre-test and from 1 to 7 in the post-test. This experience was carried out in a public primary school, located in the municipality of Padua (Italy), in a multicultural neighborhood, close to the city center.

Results showed that pupils from both classes improved significantly their knowledge of the names of family relations after the intervention as compared to before the intervention. However, the experimental group did not reach a better level of knowledge after the intervention with Pepper as compared to the control group who experienced traditional classes. The results are discussed in terms of previous studies, pointing to the short duration of intervention as a possible moderator of the little effect of the experience on learning outcomes. As for the strong points of the experience, it is worth to mention the interdisciplinary nature of the research team, who proficiently managed to "speak the same language" despite their different backgrounds.

Keywords
Educational Robotics, robot-assisted language learning (RALL), English as a second language, quasi-experimental design, language proficiency

1. Introduction

This is an exploratory work involving the use of the Pepper robot for English as-a-second-language learning in an Italian primary school. The project on robot-assisted language learning (RALL) was devised by a multidisciplinary team of engineers, psychologists and teachers, through the collaboration between the Engineering Department of the University of Padua (DEI), University of Bergamo and the 3° Istituto Comprensivo di Padova.

In the last seven years Italian schools have tried different ways to develop digital curricula, sometimes offering spotted workshops to students held by external experts, sometimes trying more systematic experiences conducted by a technology expert identified among the school's staff.

The two key documents that contributed to benchmark this interest in Italian schools are the Piano Nazionale Scuola Digitale in 2015 and the European Digital Competence Framework for Citizen (also known as DigComp) in 2017 (updated in 2022; Vuorikari et al., 2022). The first has given importance in updating the widespread school curricula, encouraging schools to enhance students' digital awareness, building makerspaces and to renovate school's libraries and laboratories. Substantial funds have been provided by the Italian Education Ministry. The latter constitutes a tool, for member states, to drive policy aimed to innovate education, thinking of the challenges that the future European citizens will face.

Robotics is part of this effort to bring innovation to Italian schools. Educational robotics (ER) can increase students' learning motivation as well as improve students' learning outcomes (e.g., Benitti, 2012; Pivetti et al., 2020; Xia & Zhong, 2018). In his review, Mubin et al. (2013) identified four major domain or subject activity for the use of social robots in public schools; one of them is the teaching of English as a Second language (ESL) as well as other languages.

As for language learning, the review by van den Berghe et al. (2019) considered a total of 33 studies and summarized the language learning results across age groups. Studies on school-aged children (e.g., Alemi et al., 2014) suggested that RALL benefits word learning more with these groups than with preschool children. Moreover, they found mixed results regarding the robot's effectiveness for word learning. Whereas several studies found only small (Movellan et al., 2009) or no learning gains (Gordon et al., 2016), other small-scale studies with preschool children showed positive effects of the use of robots in word learning and suggested that aspects such as learning by teaching and gestures might improve learning gains (e.g., de Wit et al., 2018).

The meta-analysis by Wu and Li (2024) calculated the effect size of 47 independent studies involving robot-assisted language learning (RALL) with a total of 1791 participants. The overall effect size was moderate and significant, suggesting that RALL outperforms non-RALL conditions. An analysis of possible moderators showed that educational level was a significant moderator: RALL was effective for learners of the primary and secondary education levels, rather than those of tertiary level.

Those results are also in line with another meta-analysis by Lee and Lee (2019), showing that RALL had a positive treatment effects on language learning when compared to non-RALL conditions, that is traditional teacher-only learning without technology or other technology-mediated conditions (i.e., PCs or tablets). Moreover, regarding moderators, RALL showed a larger effect size for kindergartners and upper elementary learners than for adult students. In addition, RALL had a larger average effect size for teaching vocabulary than other language aspects (i.e., listening, reading and speaking).

Tanaka et al. (2015) described a case report where they offered a scenario where children learnt together with Pepper at their home environments from a human teacher who gives a lesson from a remote classroom. Buchem et al. (2022) presented an instructional design and programming of a game-based learning scenario for learning English grammar in higher education, with Pepper. Their pilot study revealed that pupils enjoyed playing the grammar game with the Pepper robot, and perceived the robot as a friendly, kind, pleasant, funny and relaxed game partner which appeared good-humoured, jocular, funny, relaxed and interested. However, both papers only reported qualitative reflections on the experience and did not carry out an outcome evaluation in terms of children learning objectives.

This paper aims to report the experience of using Pepper to teach English words in a public primary school and to provide a rigorous outcome evaluation of the experience, in terms of pupils' learning outcome, via a pre-post test design, with a control group of pupils learning the same words through traditional classrooms. This research design would allow us to determine whether the experimental and control groups were similar in terms of knowledge of words prior to the interaction with Pepper and whether learning with Pepper would be more effective in teaching English words than human teachers.

We formulate the following hypotheses:

H_1. At the baseline in the pre-test before the intervention, pupils from experimental and control group would be similar in terms of level of knowledge of names of family relationships; in other words, the pupils of experimental and control group would be equal as for how many names of family relationship they knew before the intervention;

H_2. At the pre-post test comparison, pupils from either the experimental and the control group would improve their knowledge of the names of family relations after the intervention; in other words, we predict that in post-test, both pupils learning with Pepper and pupils learning in a traditional classroom would improve their knowledge of names of family relationships, as compared to pre-test;

H_3. At the pre-post test comparison, there would be an interaction between group and pre-test/post-test, with pupils from experimental group to score higher than pupils from the control group in the knowledge of family names. The physical embodiment of social robots increases their value as pedagogical tools as compared to traditional classrooms and to 2D technologies, such as animated characters on computers or tablets. As social robots can express their "emotions" through body movements, gestures and facial expressions, we predict that learning with Pepper would be more interesting for children, which in turn would results in more engagement and deeper learning.

2. Method and participants

A quasi-experimental research designs can tests causal hypotheses, that is whether the treatment/intervention achieves its objectives as measured by a set of indicators. Differently than experimental design, a quasi-experimental design by definition lacks random assignment of participants to treatment or control group, as is the case for school classrooms. Quasi-experimental designs identify a comparison group that is as similar as possible to the treatment group in terms of baseline (pre-intervention) characteristics, as for instance two classrooms of the same grade and school. This way, the comparison group is able to detect what would have been the outcomes if the treatment/intervention had not been implemented. If the results show some differences between experimental and control group in the post-test in favor of the experimental group, then the intervention can be said to have caused any difference in outcomes between the treatment and comparison groups (White & Sabarwal, 2008).

This experience of learning English through robotics was carried out in a public primary school, located in the municipality of Padua (Italy), in a multicultural neighborhood, close to the city center. Two third grade classes were involved: the experimental group of 23 pupils and the control group of

19 pupils. The experimental class interacted with Pepper for a two-hour class during school time in Spring 2022. The control group followed a teacher-led frontal lesson. In the control group, the teacher used the same pictures as Pepper and orally asked the pupils the questions. Children in the control group also made a concluding drawing of the family tree. The control worked on the same topic in the same week where the experimental class interacted with Pepper.

Chi-square test showed that the two classes were similar in terms of gender distribution (chi-square(1) = .77; p=.38), age range (chi-square(2) = .86; p=.65), number of pupils with learning disabilities in each class (chi-square(1) = 1.30; p=.26). In the experimental class, there were slightly more Italian (vs foreign) mother tongue pupils than in the control group (chi-square(3) = 7.83; p=.05).

Pupils had no prior experience in programming and have never been exposed to Pepper. Pepper was programmed together by Author #1 and #2, and not by pupils.

The topic to be developed in class via ER was chosen from the school curriculum, which refers to the pre A1 level of the Common European Framework of Reference for Languages (CEFR). Pepper activities covered reading, listening and speaking skills, with family-related vocabulary. The materials used were the pre A1 Starters preparation available for free on the Cambridge website.

2.1. Description of the activities with Pepper

After identifying the topic to be developed in class (family-related vocabulary and relevant questions), Pepper was programmed with a series of tasks to interact with the children.

The materials chosen consisted of pictures and words. The Cambridge materials offer a standard vocabulary, which is required for language certification.

Pepper was turned on in front of the whole class group and it presented itself. Then the activities were carried out in small groups of four pupils in a separate room with the mediation of the research teacher (see Figure 1).

We provide a description of the activities in the following lines:
1. First activity - Vocabulary presentation: Pepper presented the target vocabulary namely the family while showing a picture.
2. Second activity - Listening comprehension: Pepper asked questions about the target language and the students were asked to select the correct options between two stimuli on the screen.
3. Third activity - Reading comprehension: Pepper showed a question on the tablet and children had to choose between two answers.

4. Fourth activity - Dialogue: Pepper asked topic related questions and the pupils answered orally.

Each activity consisted of four or five different items and kids took turns interacting with the robot. Interaction with Pepper took place mainly through the robot's tablet on which images or words were shown depending on the task. The children could answer by selecting the answer they thought was correct on the tablet. For the dialogue task, the interaction took place orally. The robot's feedback to correct responses consisted of compliments, encouraging gestures or applause. The type of feedback was provided randomly by the robot.

The teacher's main duties were double folded: on the one hand, they supported the children with the reading task by reading aloud or correcting pronunciation; on the other hand, they were overseeing all the technical functioning of the robot.

The techniques used were modelling and drilling. It is important to highlight how students' motivation was enhanced by interacting with Pepper. In addition, the need to answer clearly in order to be understood by the robot led students to engage in the production of answers while trying to be precise.

The control group followed a lesson guided by the teacher, who provided pictures by naming them and asking the pupils to repeat them.

The teacher used printed pictures and showed them to the students. Students as group repeated the vocabulary after the teacher.

First activity - Vocabulary presentation: The teacher used printed pictures and showed them to the students. Students as group repeated the vocabulary after the teacher.

Second activity - Listening comprehension: the teacher asked questions about the target language and the students were asked to select the correct option between two stimuli showed by the teacher.

Third activity - Reading comprehension: the teacher wrote a question on the blackboard and the children had to choose between two answers (also written by the teacher).

Fourth activity - Dialogue: Teacher asked topic related questions and the pupils answered orally.

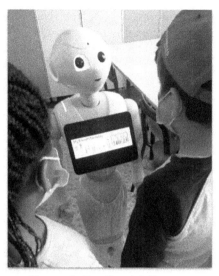

Figure 1: Pupils with Pepper

3. Programming Pepper

The experiment was carried out with a humanoid robot developed by Aldebaran United Robotics Group, called "Pepper" (Tanaka, 2015). The choice of Pepper was due to many factors. First of all, the design is very friendly and empathic compared to other available robots. Moreover, Pepper is equipped with many sensors and tools like cameras, microphones and a touchscreen tablet which make the interaction easier.

To program the robot, we relied on the APIs included in the software developer kit of Pepper, like speech recognition, speaking and motion tools inside the Choregraphe environment (version 2.5.10). We applied block-based programming where each robot's behaviour is modelled with one or more blocks containing a portion of Python (version 2.7) code editable by the programmer. We employed two main kinds of behaviours: (1) Autonomous Life Behaviors (ALB) that are meant to mimic human behaviors to make the robot look alive (e.g., the robot can look towards the person who is speaking) and (2) Target Behaviors (TB) that are instead created to target specific applications, like supporting teaching in our case (see Figure 2).

Each TB implements some activities co-designed with the teachers, including description of a picture, listening, speaking, reading and interaction activities. All the activities have been implemented with TBs. The provided library blocks, in fact, were not powerful enough to provide all the requirements. For example, the touch-to-answer mechanism is based on TBs written in Python. To give more variability, we also recorded some new actions using the technique of "learning-by-demonstration", where a human can store a set of joints positions (i.e., a motion) just by forcing the robot to execute the movement with the motors not live. Multimedia material to support the exercises, such as photos to be shown on the tablet, have been produced also through a vectorial graphic application, following the suggestions of the teachers.

We also developed some behaviours in order to give feedback on the child's answer. These behaviours include both verbal communication and gestures taken from a random pool to make the interaction with the robot more natural. Positive and negative feedback are also supposed to boost the engagement of children.

We provide a description of customizes blocks in the following lines:

Description: Pepper shows an image on the tablet and describes it to the children to revise the vocabulary of the topic (family and people in our case).

Listening: Pepper shows an image with two alternatives and asks a voice question. The child listens and touches the right one (if wrong, s/he tries again).

Speaking: Pepper shows the image of a person and asks to recognize it. The child replies verbally (if wrong, s/he tries again). If Pepper fails to understand well, the teacher can intervene with a swipe on the tablet.

Reading: Pepper shows a question with two alternatives written on the tablet screen. The child reads the alternatives and touches the answer. If wrong, s/he tries again.

Interaction: Dialogue with Pepper who asks the child to talk about his family with simple targeted questions. The child responds verbally. Different reactions are forecasted, depending on the children's answer. If Pepper fails to understand well, the teacher intervenes with a swipe on the tablet.

The whole program structure was designed to be modular and easily interpretable. Each TB implements a specific function and complex behaviors are obtained by serializing and nesting different TBs and/or predefined blocks. A system to synchronize TB and ALB have also been developed.

A simple but effective user interface (based on tablet touchscreen and voice commands) is provided. In this way, the robot is easily usable by anyone. No prior knowledge in robotics is required, neither to connect a PC to the robot. The system works as it is, just turning on the robot.

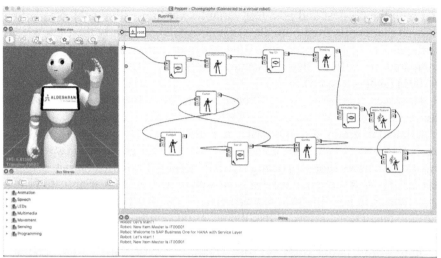

Figure 2: Example of a Pepper's application, developed in the Choregraphe environment

4. Results

A pre-experimental pre-test/post-test design with a control group was used to evaluate the outcome of the experience, in terms of pupils learning outcome. Pupils from both classes took the same test before and after the intervention. The written test covered the topic of names of family relations (e.g. uncle, grandmother etc). The test was prepared by the teachers, who relied on their experience in preparing a test targeted to the school-age and performance level of the involved pupils. The test was composed of 7 images depicting a family relation and asking pupils to name each of them by writing the name in English under the image. The final score ranged from 0 to 5 in the pre-test and from 1 to 7 in the post-test.

At the baseline in the pre-test, that is before the intervention, pupils from experimental and control groups were similar in terms of level of knowledge of names of family relationships (supporting H_1). The chi-square test showed that the pre-test score measuring the level of pre-test knowledge for each group was similar (chi-square(5)=4.44, p=.49). Additionally, as the inspection of histogram showed that the distributions of the variables were normal, we also ran a t-test on independent samples to search for possible differences between experimental and control groups. T-test confirmed the results of the chi-square test, showing that the two groups were similar before the intervention in terms of preliminary knowledge of family relations (t(40)=1.40, p=.17).

As for the outcome evaluation in terms of language learning, we compared pre-test and post-test scores of the experimental and control groups via repeated measures ANOVA, with the type of group as independent variable and with the pre-test and post-test variables varying within subjects. Results showed that pupils from both classes improved significantly their knowledge of the names of family relations after the intervention (M=4.35, SD=.28) as compared to before the intervention (M=2.46, SD=.23) ($F(1,38)$=55.15; $p<.001$) (supporting H_2).

However, there was no significant interaction between the group and the pre-test/post-test scores ($F(1,38)$=1.58; p=.22), indicating that the experimental group did not reach a better level of knowledge after the intervention with Pepper as compared to the control group who experienced traditional classes (not supporting H_3). Additionally, we also computed a post-test less pre-test score and we ran a chi-square test on this variable, to further investigate the possible differences between the two groups. Unfortunately, the chi-square test confirmed that there was no significant difference between the two classes before and after the intervention (chi-square(7)=5.53; p=.60). In other words, in our study we found that the pupils in the RALL condition did not outperform the pupils in the traditional teacher condition in terms of English as a foreign-language-learning.

5. Discussion and conclusions

Our study investigated whether primary school pupils involved in a RALL activities with Pepper would outperform their peers in traditional classes in learning English family-related words. We did this via a pre-post test study, with experimental (N=23 pupils) and control group (N=19 pupils) from third grade attending a public school in North Italy.

According to our main hypothesis (H_3), we expected to measure a better improvement for pupils in the intervention group with Pepper as compared to the control group with traditional teachers. However, our results showed improvements for both groups, the one interacting with Pepper and the one experiencing traditional class. Three studies found similar unsatisfactory results when investigating how effective robots were in teaching words in comparison to human teachers (Kory Westlund et al., 2015; Mazzoni & Benvenuti, 2015; van den Berghe et al., 2018). Our results are also inconsistent with a recent meta-analysis by Lee and Lee (2019), showing that RALL had a positive treatment effects on language learning when compared to non-RALL conditions, that is traditional teacher-only learning without technology, especially at primary school level.

One possible explanation relies on the short interaction with Pepper, being only a 2-hour class, and in the pilot nature of the experience. As for the duration of intervention, the meta-analysis by Wu and Lee (2024) found that longer durations are favored, as both intermediate (between 1 and 4 weeks) and long durations (longer than 4 weeks) achieved large effect sizes, while short duration (less than one week) had a small effect size. This could be due to the need of pupils to get acquainted with educational technologies during the first interactions with the robot and need to overcome the RALL novelty effect that could distract the pupils from focusing on the task.

Among the possible moderator of the effect of RALL on pupils, labor division refers to the distribution of duties among teachers and robots. Three categories are detected in literature: (a) no teacher, that is no teacher available in RALL, (b) assistant teacher, assisting robot to teach a language and (c) tutor, teacher-led RALL lesson. As for the teacher role, Wu and Li (2024) found that tutor achieved a large effect size, no teachers had a medium effect, while the effect of assistant was not observed. However, our results showed that Pepper, who played the role of tutor, did not outperform traditional classes.

Also, the pupils were enthusiastic to interact with Pepper and participant observation showed that student engagement was high during the class (LeTendre & Gray, 2023). Also, turn-taking during the interaction allowed every pupil to interact with Pepper, exploiting this opportunity.

As for the strong points of the experience, it is worth to mention the interdisciplinary nature of the research team, composed of primary school teachers and post-doc and faculty members from engineering and social sciences, who proficiently managed to "speak the same language" despite their different backgrounds.

We consider this experience as a valuable pilot study to test both the tools and the methodology involved. Additionally, our paper is among the few papers not only describing the experience, but also addressing the issue of outcome evaluation of our experience, in terms of pupils' learning objectives. The idea that children may be able to learn equally well when being taught by a robot or by a human teacher, or when being assisted by a robot or child peer deserves further investigation. Future studies should consider to expand the duration of the interaction with Pepper from one to four weeks, to overcome the novelty effect and allow pupils to adjust to the new technology. Moreover, the role of the teacher should be object of further reasoning as literature showed that different educational approach could lead to different learning outcome (e.g., Laurillard, 2013).

References

Agatolio, F., Moro, M., Menegatti, E., & Pivetti, M. (2019). A critical reflection on the expectations about the impact of educational robotics on problem solving capability. In Intelligent Autonomous Systems 15: Proceedings of the 15th International Conference IAS-15 (pp. 877-888). Springer International Publishing.

Alemi, M., Meghdari, A., & Ghazisaedy, M. (2014). Employing humanoid robots for teaching English language in Iranian junior high-schools. International Journal of Humanoid Robotics, 11(03), 1450022.

Benitti F. (2012). Exploring the educational potential of robotics in schools: A systematic review. Computers & Education, 58(3), 978-988

Buchem, I., Mc Elroy, A., & Tutul, R. (2022). Designing and programming game-based learning with humanoid robots: a case study of the multimodal "make or do" english grammar game with the pepper robot. In iceri2022 proceedings (pp. 1537-1545).

de Wit, J., Schodde, T., Willemsen, B., Bergmann, K., De Haas, M., Kopp, S., Krahmer, E., & Vogt, P. (2018, February). The effect of a robot's gestures and adaptive tutoring on children's acquisition of second language vocabularies. In Proceedings of the 2018 ACM/IEEE international conference on human-robot interaction (pp. 50-58).

European Digital Competence Framework for Citizen (DigComp, 2017/2022). https://joint-research-centre.ec.europa.eu/digcomp/digcomp-framework_en

Gordon, G., Spaulding, S., Westlund, J. K., Lee, J. J., Plummer, L., Martinez, M., Das, M., & Breazeal, C. (2016, March). Affective personalization of a social robot tutor for children's second language skills. In Proceedings of the AAAI conference on artificial intelligence (Vol. 30, No. 1).

Laurillard, D. (2013). Teaching as a design science: Building pedagogical patterns for learning and technology. Routledge.

Leatherdale, S. T. (2019). Natural experiment methodology for research: a review of how different methods can support real-world research. International Journal of Social Research Methodology, 22(1), 19-35.

Lee, H., & Lee, J. H. (2022). The effects of robot-assisted language learning: A meta-analysis. Educational Research Review, 35, 100425.

LeTendre, G. K., & Gray, R. (2023). Social robots in a project-based learning environment: Adolescent understanding of robot–human interactions. Journal of Computer Assisted Learning, 1–13.

Mazzoni E., Benvenuti M. (2015). A Robot-partner for preschool children learning English using socio-cognitive conflict. Journal of Educational Technology & Society, 18, 474–485.

Movellan J. R., Eckhardt M., Virnes M., Rodriguez A. (2009). Sociable robot improves toddler vocabulary skills. In Proceedings of the 4th ACM/IEEE international conference on Human robot interaction (pp. 307–308). New York, NY: ACM.

Mubin, O., Stevens, C. J., Shahid, S., Al Mahmud, A., & Dong, J. J. (2013). A review of the applicability of robots in education. Journal of Technology in Education and Learning, 1(209-0015), 13.

Piano Nazionale Scuola Digitale (2015). https://www.miur.gov.it/scuola-digitale.

Pivetti, M., Di Battista, S., Agatolio, F., Simaku, B., Moro, M., & Menegatti, E. (2020). Educational Robotics for children with neurodevelopmental disorders: A systematic review. Heliyon, 6(10).

Tanaka, F., Isshiki, K., Takahashi, F., Uekusa, M., Sei, R., & Hayashi, K. (2015, November). Pepper learns together with children: Development of an educational application. In 2015 IEEE-RAS 15th International Conference on Humanoid Robots (Humanoids) (pp. 270-275). IEEE.

van den Berghe R., van der Ven S., Verhagen J., Oudgenoeg-Paz O., Papadopoulos F., Leseman P. (2018). Investigating the effects of a robot peer on L2 word learning. In Companion of the 2018 ACM/IEEE International Conference on Human-Robot Interaction (pp. 267–268). New York, NY: ACM.

van den Berghe, R., Verhagen, J., Oudgenoeg-Paz, O., Van der Ven, S., & Leseman, P. (2019). Social robots for language learning: A review. Review of Educational Research, 89(2), 259-295.

Vuorikari, R., Kluzer, S. and Punie, Y., DigComp 2.2: The Digital Competence Framework for Citizens - With new examples of knowledge, skills and attitudes, EUR 31006 EN, Publications Office of the European Union, Luxembourg, 2022, ISBN 978-92-76-48883-5

Westlund, J. K., Dickens, L., Jeong, S., Harris, P., DeSteno, D., & Breazeal, C. (2015, October). A comparison of children learning new words from robots, tablets, & people. In Proceedings of the 1st international conference on social robots in therapy and education.

White, H., & Sabarwal, S. (2008). Quasi-experimental design and methods, methodological briefs: Impact evaluation 8. Florence, Italy: UNICEF Office of Research

Wu, X., & Li, R. (2024). Effects of Robot-Assisted Language Learning on English-as-a-Foreign-Language Skill Development. Journal of Educational Computing Research, 07356331231226171.

Xia, L., & Zhong, B. (2018). A systematic review on teaching and learning robotics content knowledge in K-12. Computers & Education, 127, 267-282.

Engaging children with social robot-led Nutritional Education

Loredana **Perla**[1], Berardina **De Carolis**[2] and Stefania **Massaro**[3]

[1] *University of Bari, P.zza Umberto I 1, Bari, 70100, Italy, 0000-0003-1520-0884, loredana.perla@uniba.it*
[2] *University of Bari, P.zza Umberto I 1, Bari, 70100, Italy, 0000-0002-2689-137X, berardina.decarolis@uniba.it*
[3] *University of Bari, P.zza Umberto I 1, Bari, 70100, Italy, 0000-0003-4695-1007, stefania.massaro@uniba.it*

Abstract

Introduction: Social robots have shown potential as engaging educational agents due to their physical embodiment and ability to promote social behaviors. Previous research has shown that robots can improve student motivation and engagement in primary school educational settings. This paper presents an exploratory research project on social robot-based obesity nutritional education in a school setting. The primary aim of this study is to investigate the effectiveness of using the robot Pepper acting as a teacher to promote healthy nutrition and prevent overweight and obesity in primary school children.

Study design: a school-based pilot intervention was carried out on with 34 fourth-grade children (9-10 years old) divided into two groups. The experimental group attended a lesson on the food pyramid with Pepper as the teacher and participated in storytelling and game-based activities on its tablet, while the control group underwent the same activities with a human teacher.

Methods: The intervention's efficacy was evaluated by administering questionnaires to participants at the beginning and at the end of the activities to measure changes on children's knowledge and engagement.

Results: Results suggest that Pepper outperformed the human teacher in terms of engagement, as perceived by the participants. Additionally, both groups demonstrated an increase in knowledge about healthy nutrition following the intervention.

Conclusions: These findings indicate that social robots, show potential as effective educational agents on nutritional education in school settings. Longer interventions are needed to fully understand the potential of social robots in the framework of a nutritional school-based curriculum.

Keywords

Social Robotics, Educational Robotics, Engagement, Obesity
Prevention, Nutritional Education

1. State of the art

Childhood obesity represents a major health problem affecting educational attainment and quality of life with psychological correlates such as depressive and anxious symptoms and low self-esteem (WHO, 2022). In children, obesity has been increasingly linked to a spectrum of non-communicable diseases as diabetes, cardiovascular diseases, and cancer. Studies highlight the critical importance of addressing childhood obesity against long-term health complications (Perla & Massaro, 2022). In this respect, European Commission (2022) asks for participatory school-based interventions on nutrition education to promote capacities of informed decision making and critical thinking.

The idea of using social robots for teaching and learning at school has become increasingly relevant (Ekström & Pareto, 2022). Current research focuses on the design of school-based experimental initiatives with social robots, based on the functionalities that enable these robotic agents to promote social interaction and create relationships with students (Woo, LeTendre, Pham-Shouse, & Xiong, 2021). The deployment of these technologies in educational settings is justified by evidence on the enhancement of students' motivation, trust, and engagement, which are considered to be crucial elements in the process of successful learning (Stower, Calvo-Barajas, Castellano, & Kappas, 2021). Social robots possess the ability to captivate students' attention employing interactive features and personalised learning and creating an immersive learning environment.

Social robots are artificial intelligence platforms, paired with sensors, cameras, microphones and computer vision technology. These robotic agents represent an embodied artificial intelligence able to display appropriate, responsive, and adaptive behaviors (Breazeal, Dautenhahn, & Kanda, 2016). They interact with people in everyday environments, following social behavior typical of humans: they can interpret human behavior properly, react to changes during the interaction, make decisions, behave in a socially plausible manner, and learn from a user's feedback and previous interactions (Malerba et al., 2019; Perla, De Carolis, Massaro, & Vinci, 2023).

Different purposes of the usage of social robots in education are:
- effectiveness: to support knowledge and skill acquisition;
- engagement: to make children more involved in learning activities;
- special needs: to support learners with specific difficulties;
- empowerment of young patients on healthy lifestyles;
- language learning: to support vocabulary learning.

The role and impact of social robots in education have been specifically investigated by Belpaeme, Kennedy, Ramachandran, Scassellati, and Tanaka (2018). The potential roles of social robots in education have been examined, with particular focus on the roles of tutors and peers. The robot can be used as a peer, in the form of an experienced peer who helps the learner as a fellow student, or even a peer who needs help. In the latter case, the student becomes a teacher for the robot-peer (learning by teaching), while in the former the robot-peer indirectly assumes the role of teacher. In recent years, a prevalent approach to the utilization of robots in education has been that of tutors. The role of a robot-tutor is to teach to a single pupil or to a small group. The subjects that can be taught by robots encompass the teaching of science subjects such as mathematics or physics, as well as the promotion of psychosocial skills. Robot-led lessons frequently comprise brief sessions during which responses to questions are employed to facilitate more tailored responses to children's individual knowledge. The strategies employed in robot-based tutoring scenarios encompass classroom discussion, storytelling, scaffolding (Wood et al., 1976), and game-based learning. Game-based learning is an active learning strategy that encourages active participation and learning outcomes through the use of interactive games and simulations to engage students in their learning experiences. This approach has been shown to foster children's holistic development (Plass, Mayer, & Homer, 2020). Robots are traditionally deployed in games as a means of exploring human-robot interactions (Rato, Correia, Pereira, & Prada, 2023).

Research has provided preliminary evidence of the efficacy of social robots on educational nutrition and childhood obesity prevention. Social robots have emerged as an innovative tool for teaching healthy nutrition engaging students in learning about the importance of a balanced diet and the benefits of healthy eating habits. A study on the use of socially assistive robotic technologies for educational interventions on healthy food choices with children demonstrated a high level of enjoyment when interacting with the robot and a statistically significant increase in engagement over the duration of the interaction. Furthermore, evidence of the establishment of a relationship between the child and the robot and encouraging trends towards learning were found (Short et al., 2014). A social robot-based platform was found to be an interesting and useful tool for motivation and guidance of children towards the achievement of personal behavioral goals, thereby proving its acceptability and feasibility (Triantafyllidis et al., 2022). The study by Rosi et al. (2020) on the social robot Nao, specifically used in teaching activities to explain nutritional concepts and actively participate in an educational game, demonstrated no discernible impact on children's knowledge. However, due to the clear ability of NAO to increase the curiosity of the children, it highlights the need for longer interventions to fully understand the potential of this social robot in the context of a nutritional-based school curriculum. The Alpha Mini humanoid robot has been tested as a nutritional coach, utilizing personalized strategies to enhance awareness and motivation in children through dialogue and serious games. The results demonstrated a positive impact on user experience, motivational strength, and trust (Abbatecola, De Carolis, & Oranger, 2022). Therefore, further interventions are needed to fully understand the impact and effectiveness of social robots on this educational context.

Theoretical perspectives support the integration of social robots in educational settings, emphasising the potential of robots to enhance learning experiences through technology. A key theoretical approach is that of constructivism, which posits that learned knowledge is shaped by what the learners know and experience. Another theoretical perspective is that of constructionism, which states that learning occurs when a student constructs a physical artefact and reflects on his/her problem-solving experience based on the motivation to build the artefact. A further theoretical perspective is that of social constructivism, which, drawing on Vygotsky (1978), emphasizes the role of the community in knowledge (Mubin, Stevens, Shadid, Al Mahmud, & Dong, 2013).

The theory of enaction proposed by Varela, Thompson, and Rosch (1991) presents a comprehensive view of cognitive science and human experience, advocating for a more holistic understanding of the mind-body relationship as a dynamic process linked to bodily identity and situated within a field of relationships. The concept of an embodied mind underscores the influence of the body on the mind and the interconnectedness between bodily experiences and cognitive processes. The enactive framework encourages the consideration of all dimensions of experience in research, integrating the interior and exterior aspects of cognition. Social robots, as embodied agents, enrich interactions by acting as mediators between students, teachers and knowledge, potentially empowering students.

2. Introduction

This transdisciplinary research, connecting education and computer science, aims at exploring the potential of using a robot as a tutor to engage students on nutritional knowledge and innovate nutritional education, considering nutritional education a subject where teachers usually perceive multiple teaching-learning barriers (Jones & Zidenberg-Cherr, 2015). The overall goal is to create an entertaining and motivational learning environment to involve children in the process of reconsidering their health behaviors specifically targeting diet behaviors. We set out two research questions: (1) to what extent does incorporating a social robot into a school-based nutritional education program impact children's level of engagement? (2) Can the child-robot interaction improve learning outcomes on this subject?

Engagement plays a pivotal role in technology-enhanced learning research. It sustains interest, participation, and involvement during the learning process. Enhancing sustained student engagement has become a critical educational objective for two key reasons: engagement is a prerequisite to meaningful learning, and maintaining engagement involves cognitive and socio-emotional skills that are learning objectives in themselves (OECD, 2021).

The results of a within-subjects field experiment that compared the social robot Pepper acting as a teacher to a human teacher are presented. We used the social robot Pepper, a 1.2-m-tall, wheeled humanoid robot, capable of exhibiting body language, interacting and moving around. It can analyze people's expressions and voice tones, as it is equipped for multimodal communication with the humans around it. It is a carefully shaped robot, without any sharp edges, for a more appealing and safer presence in the human environment and it is equipped with a tablet. Its anthropometric characteristics allows it to be perceived more as learning companion rather than an artificial toy (Pandey & Gelin, 2018).

The theoretical framework of this research is represented by the cultural and ecological perspective of human development and learning, which is understood as a the result of the interaction of cognitive, social, and affective factors (Vygotsky, 1978). This work is grounded in the theory of embodied cognition, which supports the use of educational robots to facilitate learning through intentional, meaningful, and socially embedded interactions, to foster cognitive development, positive emotional states, and effective learning outcomes. Educational robots promote a high level of involvement in the experience, social interactions, and positive emotional states, which can shape attitudes and generate fun and learning (Hoffmann & Pfeifer, 2018).

3. Methods

3.1. Participants

The study comprised a total of 34 fourth-grade school children (9-10 years old), from two classes of a private school in Italy (Table 1). This age group was selected due to their cognitive capability to complete questionnaires for evaluation purposes and considering that obesity often emerges around this age. Participants were recruited based on the following inclusion criteria:

- age: participants were required to be within the age range of 9-10 years old
- grade: participants must be enrolled in the fourth grade
- ability to understand and complete questionnaires independently.

Table 1

Demographic characteristics of participants

Characteristics	Experimental Group (n=20)	Control Group (n=14)	Total (N=34)
Age (years)	Mean ± SD: 9.4 ± 0.3	Mean ± SD: 9.5 ±0.4	Mean ± SD: 9.4 ±0.4
Gender:			
- Male	10 (50%)	7 (50%)	17 (50%)
- Female	10 (50%)	7 (50%)	17 (50%)
Educational background:			
Private school	20 (100%)	14 (100%)	34 (100%)

Note: SD = Standard Deviation

Participants were randomly assigned to either the experimental group or the control group: 20 children were in the experimental group and 14 in the control group. Both groups had similar mean ages, with the experimental group having a mean age of 9.4 years (SD=0.3) and the control group having a mean age of 9.5 years (SD=0.4). The distribution of genders was evenly split in both groups, with an equal number of males and females.

3.2. Design

We carried out an exploratory study with two independent groups. The two intervention groups were exposed to a nutritional educational lesson, a storytelling activity, and a playful phase (4 hours). The experimental group had the robot as a teacher while the control group, underwent the same activities with a human teacher.

The hypothesis formulated in the study was that the two experimental conditions (robot-teacher and human-teacher) would produce different results in terms of:

1. learning: variations were expected in learning outcomes between the two instructional conditions, specifically related to understanding the importance of the nutritional pyramid.
2. engagement: differences were expected in how actively and attentively the fourth-grade school children participated during the lesson when taught by the robot Pepper compared to when taught by a human teacher.

Engagement has been used to describe diverse behaviors, thoughts, perceptions, feelings, and attitudes, but researchers generally agree that engagement, as a goal-directed state of active and focused involvement, is a multidimensional construct, constituted by *emotional, behavioral and cognitive* components (OECD, 2021).

A human facilitator was present to provide support and guidance in the robot session. The presence of a human facilitator during the robot session was considered as a potential confounding variable. The facilitator's behavior support or guidance might differ between the robot and human conditions. To address the potential confounding variable of the human facilitator, strategy of standardized facilitator training was employed: training the human facilitator to provide similar support in both the robot and human sessions and ensure that the facilitator's behavior does not favor one condition over the other.

3.3. Procedure

The intervention was divided into three sessions: lesson, storytelling and playing. During each session, the teacher (R and H) introduced the topic and illustrated the nutritional pyramid. - In the experimental group, the robot presented the lesson using verbal interactions and visual support (Figure 1) while in the control group, the human teacher presented the lesson using traditional methods such as verbal explanation and slides.

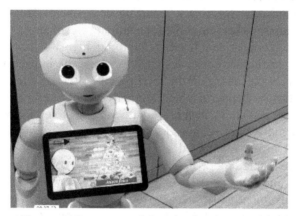

Figure 1: Pepper explains the food pyramid

Students were involved in a storytelling activity, traditionally considered as a powerful teaching tool to promote cognitive, social, and emotional development in students, on healthy eating. The story aims to encourage children to try foods even if they look unfamiliar and emphasizes the benefits of whole versus processed food. The stories include a concluding comprehension question to evaluate children's understanding. Responses can be given through vocal input or by selecting options on the tablet. Pepper reinforces correct answers with advice and corrects wrong answers explaining the right choice and providing guidance. Eventually every group of participants was invited to play games on food identification on Pepper's tablet in order to provide enjoyment to all the students involved. The selected games, adapting the app Pepperhero, have been 'The traffic light 'game, reflecting the colors of the food pyramid. In this activity, where children were challenged to place foods in the predetermined colors: green, yellow, and red. Other included games were 'True or false', featuring questions related to the food pyramid, and a quiz resembling "Who Wants to Be a Millionaire" that encouraged selecting correct answers.

The activity comprised a pre-test and post-test phase. Prior to and one week after the intervention, the human teachers in each class administered a self-administered questionnaire. The teachers provided instructions on completing the questionnaire and remained in the classrooms to address any questions or requests for clarification. The control group underwent assessment concurrently (Table 2). The administration of pre-questionnaires played a crucial role in analyzing students' skills in food education. The primary objective was to assess students' familiarity and awareness of food-related topics. Questions encompassed a range of concepts, from basic knowledge to more in-depth reflections, and covered various facets of food education. Topics included knowledge of daily portions of fruits and vegetables, with multiple-choice responses. Furthermore, the students' sporting activities were investigated in order to assess their habits in relation to a healthy lifestyle, with food education forming a recurring component. This initial phase of preliminary analysis served as the foundation for evaluating the effectiveness of the proposed educational intervention through post-activity questionnaires. A post-test questionnaire was designed to assess the impact of the lesson on nutrition education, based on the food pyramid, the storytelling activity and game activities on Pepper's tablet. Students were tasked with identifying the macronutrients present in each level, with the objective of providing concrete examples or more practical questions on daily consumption. In this study the experimental group conducted an assessment to evaluate various aspects related to the social robot Pepper. Specifically, participants in R's group responded to questions aimed to explore their views on Pepper's acceptability, level of involvement, enjoyment during interactions, and perceived efficacy as an educational tool (Table 3) while, the group taught by the human teacher received a questionnaire consisted of 4 multiple-choice questions specifically focused on Pepper's perception.

Table 2
Pre-test and post-test questionnaires

Pre-test		
All children	On personal knowledge and beliefs on healthy eatings	20 multiple choice questions
Post -test		
All children	Knowledge on nutritional pyramid	7 multiple choice questions
R's group	Pepper's acceptability, involvement, enjoyment and perceived efficacy	12 questions with a 5-point- scale from 'strongly agree' to 'strongly disagree'
H's group	Pepper's perception	4 multiple-choice questions

Table 3
Questions from post-test phase with R's group

If I say..	You completely agree	You agree	You don't agree	You disagree	You strongly disagree
"Pepper helps you in school to learn new "things," you..					
"Pepper helped me understand what foods can help me eat healthier," you..					
"I wish I could use a robot like Pepper in school," you..					
Did you have fun with Pepper in the classroom?					

171

4. Results

Data collected from students were analysed with software SPSS® using t-test to compare the averages between the two groups and the Pearson index to measure the relationship between different variables and obtain complementary information.

An in-depth comparison was made between the results of the group run by the human teacher and those of the group involved with Pepper. In the pre-test data, the teacher's class recorded a higher average score than the class in which Pepper subsequently intervened. However, in the post-test, the dynamic was reversed, showing a minimum deviation of 0.2 tenths at the average level. This change in post-test performance could indicate a positive impact of Pepper's presence.

A correlation analysis of the use of Pepper at school was conducted based on another questionnaire proposed on the day of the experience. This was done in order to understand the impact of Pepper within the school environment in an objective manner. From this experimental study of Pepper as a teaching tool in school, several correlations emerged between students' perceptions and interaction with the robot. The thresholds adopted to define positive (>0.33) and negative (<-0.33) correlations were taken as reference criteria.

1. General correlations:
   - significant positive correlations (>0.33) were found between the concept of Pepper as a tool for learning new things, the engagement felt during the lesson and the trust in Pepper as a teacher.
   - a less strong positive correlation (0.35) was found between Pepper's perceived usefulness at school and enjoyment, as well as engagement and trust.
2. Engagement and Fun:
   - Engagement with Pepper as a teacher was positively correlated with the fun experienced during the lesson. Moreover, engagement and fun are strongly positively correlated (0.64).
   - The test of primary school children of the same age confirmed the positive correlation between trust and fun (0.5).
3. Perception of Pepper as an intruder:
   - An inverse correlation was found between the perception of Pepper as an intruder and both enjoyment and engagement during the lesson.
4. Promotion of Healthy Eating:
   - A positive correlation was observed between the perception of Pepper as a promoter of healthy eating and involvement during the lesson. This indicates that those who felt involved also perceived Pepper as a promoter of healthy eating habits.

Observations during the activities shoed that children engaged with Pepper in various ways during nutritional education sessions. They demonstrated curiosity, exploring Pepper's features and asking numerous questions about its origins and preferences. Pepper's presence fostered discussions and playful interactions. Peer collaboration was facilitated by Pepper, with children discussing food groups and sharing tips. Through these interactions, Pepper evolved from a mere machine to a cherished companion in the eyes of the children. Data collected from a group interview with the teacher and the school dean expressed significant interest in the study's findings, particularly regarding the engagement of children with autism during the activities. They observed that Pepper, the robot-teacher, elicited positive responses from these students, leading to increased participation and interaction. This positive reaction was seen as a promising development in leveraging technology to support the educational experiences of students with autism. Overall they noticed a marked increase in students' engagement when Pepper.

5. Discussion

This exploratory study, which employs correlation analysis and pre-post data comparison, provides intriguing insights, and prompts in-depth reflections on Pepper's role as an educational tool. Initially, a significant positive correlation emerges between the perception of Pepper as an effective learning tool and the engagement felt during the lesson, indicating a close connection between interaction with the robot and the positive educational experience of the students. This association is also reflected in the trust placed in Pepper as a teacher, which suggests a potential positive impact on the perception of the robot's educational role.

A particularly noteworthy aspect of the data is the comparison between the teacher's class and the subsequent class involved with Pepper in the pre-test and post-test results. In the pre-test results, the class traditionally run by the teacher presented a higher average score, suggesting an initial effectiveness of the conventional teaching method. However, in the post-test, the average deviation of the class taught by Pepper was a minimum of 0.2 tenths, which serves to illustrate the positive potential of the innovative approach supported by technology. This suggests that although traditional methodology may excel in certain contexts, Pepper's introduction may prove particularly effective when teaching requires an in-depth focus on specific, interactive topics.

The inverse correlation between the perception of Pepper as an intruder and enjoyment, together with the positive correlation between the perception of Pepper as a promoter of healthy eating and engagement during the lesson, emphasizes the importance of an educational approach to include Pepper in the learning environment.

The experiment thus indicates that Pepper, when strategically integrated with specific topics, can enrich the educational experience, promoting engagement and learning. However, it is essential to consider students' perceptions and to customize the interaction with the robot in order to maximize the benefits. These findings suggest a new vision for innovation in education, indicating that the skillful use of technology can positively transform the learning process.

6. Limitations and Future Work

Certain limitations may have reduced the impact of the social robot on children's experience and the results of this study. Firstly, the nutritional intervention was conducted in a single educational session, during which children were presented with a considerable amount of information. Although the nutritional intervention resulted in an improvement in children's knowledge, surely a more effective learning can be achieved if the intervention is conducted on repeated occasions. A second limitation of this study is the small number of children involved in the nutritional intervention, due to the pilot nature of this study. Future implications include teacher-robot collaboration to investigate hybrid models where robots collaborate with human teachers, how can robots complement teachers' efforts, design scenarios where the robot and teacher work synergistically and explore teachers' societal acceptance and attitudes toward robot-assisted education.

The results of the study and the reflections arising from the in-depth analyses indicate that further longitudinal studies are required to examine the long-term effects of Pepper on learning and student-robot interaction. Such in-depth investigation could provide valuable information on the sustainability of the observed dynamics and the lasting effectiveness of using social robots in notional education curricula. In addition, the investigation should be expanded to encompass a range of educational contexts and levels of education. The effectiveness of Pepper may vary depending on the age of the students, the subjects covered and different teaching approaches.

Acknowledgements

This work has been funded by the Telemedicine for Obesity and Quality of Life Education - Horizon Seeds UniBa.

References

Abbatecola, A., De Carolis B., & Oranger, E. (2022). Using a Personal Social Robot as a Nutrition Coach. IUI Workshops, 222-226.

Belpaeme, T., Kennedy, J., Ramachandran, A., Scassellati, B., & Tanaka, F. (2018). Social robots for education: A review. Science robotics, 3(21), eaat5954.

Breazeal C., Dautenhahn K., Kanda T., Social robotics. Springer Handbook of Robotics, Springer (2016), pp. 1935-1972.

Catlin, D., & Blamires, M. (2010). The Principles of Educational Robotic Applications (ERA): A framework for understanding and developing educational robots and their activities. In: J. Clayson & I. Kalaš (Eds.), Constructionism 2010: Constructionist Approaches to Creative Learning, Thinking and Education: Lessons for the 21st Century: Proceedings for Constructionism 2010: The 12th EuroLogo Conference, 16-20 August, 2010 Paris, France Paris.

Ekström, S., & Pareto, L. (2022). The dual role of humanoid robots in education: As didactic tools and social actors. Education and information technologies, 27(9), 12609-12644.

European Commission, Directorate-General for International Partnerships, (2022). Action plan on nutrition seventh progress report April 2021 – March 2022, Publications Office of the European Union.

Hoffmann, M., & Pfeifer, R. (2018). Robots as Powerful Allies for the Study of Embodied Cognition from the Bottom Up. In A. Newen, L. De Bruin & Shaun Gallagher (Eds.), The Oxford Handbook of 4E Cognition. Oxford: Oxford University Press.

Jones, A. M., & Zidenberg-Cherr, S. (2015). Exploring nutrition education resources and barriers, and nutrition knowledge in teachers in California. Journal of nutrition education and behavior, 47(2), 162–169.

Malerba, D., Appice, A., Buono, P., Castellano, G., De Carolis, B., de Gemmis, M., Polignano, M., Rossano, V., & Rudd, L. M. (2019). Advanced Programming of Intelligent Social Robots. Journal of E-Learning and Knowledge Society, 15(2).

Miller, C. L., & Batsaikhan, O., Chen, Y., Pluskwik, E., & Pribyl, J. (Eds) (2021). Game-Based and Adaptive Learning Strategies. Retrieved from https://mlpp.pressbooks.pub/gamebasedlearning/.

Mubin, O., Stevens, CJ., Shadid, S., Al Mahmud, A., & Dong, JJ. (2013). A review of the applicability of robots in education. Technology for Education and Learning, 1, 1-7.

OECD (2021), OECD Digital Education Outlook 2021: Pushing the Frontiers with Artificial Intelligence, Blockchain and Robots, OECD Publishing, Paris.

Pandey, A. K., & Gelin, R. (2018). A mass-produced sociable humanoid robot: Pepper: The first machine of its kind. IEEE Robotics & Automation Magazine, 25(3), 40-48.

Perla, L., & Massaro, S. (2022). Virtual Patient Education Scenarios: Exploratory Step in the Study of Obesity Prevention Through Telemedicine. In: G. Casalino et al. (Eds.), Higher Education Learning Methodologies and Technologies Online. HELMeTO 2021. CCIS, 1542. Cham: Springer.

Perla, L., De Carolis, B., Massaro, S., & Vinci, V. (2023). Promoting Health and Wellbeing: Harnessing the Potential of Social Robots in English L2 for Elderly Cognitive Decline prevention. Helmeto 2023. 5th International Conference on Higher Education Methodologies and Technologies Online, (pp. 151-153). Roma: Studium.

Plass, J. L., Mayer, R. E., & Homer, B. D. (Eds.). (2020). Handbook of game-based learning. Mit Press.

Rato, D., Correia, F., Pereira, A., & Prada, R. (2023). Robots in Games. Int J of Soc Robotics 15, 37–57.

Rosi, A., Dall'Asta, M., Brighenti, F., Del Rio, D., Volta, E., Baroni, I., Nalin, M., Coti Zelati, M., Sanna, A., & Scazzina, F. (2016). The use of new technologies for nutritional education in primary schools: a pilot study. Public health, 140, 50–55.

Short, E., Swift-Spong, K., Greczek, J., Ramachandran, A., Litoiu, A., Grigore, E.C., Feil-Seifer, D., Shuster, S., Lee, J.J., Huang, S., Levonisova, S., Litz, S., Li, J., Ragusa, G., Spruijt-Metz, D., Mataric, M., Scassellati, B. (2014). How to train your DragonBot: socially assistive robots for teaching children about nutrition through play, 23rd IEEE International Symposium on Robot and Human Interactive Communication, IEEE, pp. 924–929.

Stower, R., Calvo-Barajas, N., Castellano, G., & Kappas, A. (2021). A Meta-analysis on Children's Trust in Social Robots. International Journal of Social Robotics, 13, 1979 - 2001.

Triantafyllidis, A., Alexiadis, A., Elmas, D., Gerovasilis, G., Votis, K., & Tzovaras, D. (2022). A social robot-based platform for health behavior change toward prevention of childhood obesity. Universal access in the information society, 1–11. Advance online publication.

Varela, F. J., Thompson, E., & Rosch, E. (1991). The Embodied Mind: Cognitive Science and Human Experience. MIT Press.

Vygotsky, L. S. (1978). Mind in Society: The Development of Higher Psychological Processes. Cambridge, MA: Harvard University Press.

Woo, H., LeTendre, G. K., Pham-Shouse, T., & Xiong, Y. (2021). The use of social robots in classrooms: A review of field-based studies. Educational Research Review, 33, Article 100388.

Wood, D., Bruner, J.S., &Ross, G. (1976). The role of tutoring in problem solving. J . Child Psychol. Psychiat., 17, 89-100. Pergamon Press. Printed in Great Britain.

World Health Organization (2022). Report on the fifth round of data collection, 2018–2020: WHO European Childhood Obesity Surveillance Initiative. Retrieved from https://www.who.int/europe/publications/i/item/WHO-EURO-2022-6594-46360-67071

Developing autonomous learning processes in Astrophysics programming with Ozobot

Laura **Leonardi**[1], Maura **Sandri**[2], Laura **Daricello**[3] and Claudia Mignone[4]

[1] *INAF Osservatorio Astronomico di Palermo, Piazza del Parlamento 1, Palermo, 90134, Italy, 0000-0001-6555-4934, laura.leonardi@inaf.it*
[2] *INAF Osservatorio di astrofisica e scienza dello spazio di Bologna, Via Piero Gobetti, 93/3, Bologna, 40129, Italy, 0000-0003-4806-5375, maura.sandri@inaf.it*
[3] *INAF Osservatorio Astronomico di Palermo, Piazza del Parlamento 1, Palermo, 90134, Italy, 0000-0002-0640-081X, laura.daricello@inaf.it*
[4] *INAF Sede Centrale, Viale del Parco Mellini 84, Rome, 00136, Italy, 0000-0002-5525-314X, claudia.mignone@inaf.it*

Abstract

This article presents an overview of recent projects, developed by the Italian National Institute for Astrophysics (INAF) as part of its third mission, illustrating some astronomy educational laboratories developed with the use of Ozobot, one of the smallest programmable robots for innovative teaching.

These new types of didactic experimentation, combining different methodologies with the support of Information and Communication Technologies (ICT), aim to teach astronomy by exploiting the potential of technology, since visual and active education allows to engage different targets of students and empower their creativity.

In addition, by developing software with Ozobot, programming with colour codes and OzoBlockly, students learn the basis of algorithms, thus improving autonomous learning processes and increasing their interest and motivation.

Lastly, the article presents results from an exploratory study observing the interaction between students and non-humanoid robots, to measure the effectiveness of Ozobot as an innovative tool to understand science, especially Astrophysics, better.

Keywords
Educational Robotics, Exoplanets, Astronomy, Astrophysics, Coding, Computational Thinking

1. Introduction

Sky observations, astronomical images, space exploration, videos and simulations of a particular phenomenon of the universe have always stimulated great fascination, especially in young people (Madsen & West, 2003). Thus astronomy, particularly through robotics and other innovative public engagement activities, can be used to encourage pursuing a scientific path of study.

Robotics is a science that embraces different disciplines such as engineering, computer programming, mechanics, and even psychology and biology. Even if it may be considered too futuristic to become an essential part of education (Bers, 2019) - to train the youngest starting from scratch, with programming and development, at all the stages of the learning process (Kelly, 2022) - it is a powerful method for the students to learn scientific subjects, increase the ability to train, solve problems and activate what is called "Computational thinking" (Wing, 2006).

In a world characterized by extraordinary technological innovations, it is necessary to have proper training in STEM subjects (science, technology, engineering and mathematics). This is useful to become an active user of technology, instead of a passive one (Montague, 2014). For this reason, it is strongly recommended to include educational robotics in the teaching program of primary schools (Chang, 2010). Educational robotics is crucial for the training of young people to allow them to be competitive in the future. Therefore, we explore the combination of robotics and astronomical content to leverage the educational potential of the former and the inspiration power of the latter.

A premise must be made: this work has been developed within the public engagement programme of INAF, the Italian Research Institute for Astronomy and Astrophysics. Our Institute coordinates research activities in the fields of astronomy and astrophysics, in collaboration with universities and other public and private entities. Under statute, INAF also promotes the dissemination of scientific results and astronomy to schools and society through public engagement activities.

We aim to contribute to quality and inclusive education, which is a basic human right, not a privilege, and should be accessible to everybody (Sustainable Development Goal 4; United Nations, 2015).

Thus, we use low-cost innovative methods and free license software to help students get closer to the scientific subjects that are always considered challenging. Exploring the abilities of educational robotics and the interaction with students from a psychological point of view is beyond the scope of this work.

2. Exoplanets and Ozobot

Activities developed with Ozobot[3] by the National Institute for Astrophysics (INAF) in this article, will help teachers to find new methods for communicating science and astronomy to the students.

In particular, the educational activities presented in this article focus on exoplanetary research, a scientific field that had a blast in the last 30 years, after the first planet beyond the Solar system was identified by Michel Mayor and Didier Queloz, who were awarded the Nobel Prize in Physics in 2019 for discovering the planet *51 Pegasi b*, in 1995 (Mayor & Queloz, 1995).

Today astronomers know more than 5000 extrasolar planets, or exoplanets, namely planets orbiting around stars other than the Sun, and public interest began to wonder: "How do stars and planets come into being? How do we study and know them?".

Using educational robots and coding, we developed an interactive journey to model the analysis techniques used by astronomers to detect the presence of planets, and to simulate the changes that take place inside the stars and in their surroundings over millions or even billions of years. These laboratories, which last each less than an hour, simulated **the method of transits** and **the method of radial velocity**. These two methods are among the most used by researchers in astronomy to find exoplanets. We recreate these two analysis techniques to provide a pedagogical resource and better explain the science behind the discoveries. Observing these planets is not easy because they are small, far and do not emit light. Thus, researchers use indirect evidence of their presence on variations in the light emitted by their host star, or on the host star's motion, reflected in the colour of its light.

We presented these two activities for the first time in 2019 at a scientific festival in Palermo (Italy) in front of about 15,000 people. The public was mainly composed of students aged 6 to 19. They approached the laboratory with great enthusiasm, recognised the power of the Ozobot-based activity in disseminating astronomical concepts and were able to correctly explain what they had experienced.

Again in 2019, in a meeting organized in Marseille (France) at the Haute-Provence Observatory, we held two workshops for teachers. The feedback was positive, more than 50 teachers and educators appreciated the way we tried to explain such a difficult subject. They considered the activities engaging and easy to understand. Moreover, teachers appreciated the fact that this small robot was cheap, so a class of students could easily buy it.

---

[3] https://ozobot.com/

2.1. First activity: the method of transits

The first activity (age 11-16) explains how *the method of transits* works. It is based on the measurement of the brightness of the star and records its periodic decreases, imperceptible to the human eye, but detectable by modern telescopes, that occur when a planet passes in front of its host star along the line of sight of observers on Earth. In fact, this is not the case for any exoplanet: its orbital plane must be aligned with our line of sight, otherwise we do not see the transit.

To explain the physics behind this method, we use the communicative potential of Ozobot, turning the robot into an exoplanet, orbiting around a star outside our Solar System. First, we use a white cardboard sheet and, in the centre of it, we draw and colour a large star: the host star of the planet Ozobot.

Then, we trace an orbit around the star, by using the markers supplied with the Ozobot Evo kit. Lastly, with the support of a torch, we simulate the light that comes from the star. Each time the planet (Ozobot) passes between the star's light and our point of observation, creating occultations, the light variations can be detected, just like they are detected in actual observations by researchers (further details in the video by scanning the QR code in Figure 1).

Figure 1: Simulation of the passage of the planet (Ozobot) in front of the star (a torch), which leads to a decrease in the brightness of the observed star, due to the obscuration induced by the planet. Credits: L. Leonardi - INAF

It is important to clarify that the observation enabled by this activity is only qualitative, not quantitative. We do not take into consideration brightness measurements, as astronomers do, but we make it possible to appreciate the decrease in brightness only, which is sufficient to understand the fundamentals of this method.

2.2.   Second activity: the radial velocity method

The second method - the radial velocity method - is based on the Doppler effect and is somewhat tougher to explain to young students. Due to the mutual nature of gravity, a planet does not orbit around its host star but both objects orbit around a common centre of mass; however, since the star's mass is much larger than the planet's, the centre of mass of the system is located within the star itself. As a result, while the planet orbits around the star, it feels the planet's pull and wobbles around the centre of mass, affected by the planet's gravitational influence. These shifts are reflected in changes in its radial velocity (the component of the star's velocity facing in the observer's direction), which can be detected by telescopes on Earth or in space in terms of a Doppler shift. Looking at the spectrum of the source, we can see that particular features in the spectrum, called lines, move towards red when the star is moving away from us, and towards blue when the star is moving towards us.

To show and explain this method, Ozobot simulates the movements of the star (and thus the changes in its spectrum), that passes from the red to the blue line and vice versa, whenever it approaches or moves away from our point of view. In actual research, astronomers record these variations, looking for the velocity variation of the star induced by the planet. For this activity, Ozobot shows the variation of colours, thanks to its colourful LEDs that change colouring to red and blue (further details in the video by scanning the QR code in Figure 2).

2.3.   Developing coding skills

After explaining the theory, both concerning educational robotics and the research for exoplanets, we proposed hands-on activities to students to make them part of the creative learning process.

Our activities sparked curiosity ("What is that? A robot? Let me see"), while students wanted to understand the science behind it ("Is it simulating a planet/star? Cool"), testifying to the strong impact that educational robotics has in teaching science. Thanks to these gimmicks, both researchers and educators could convey the scientific message to students.

But Ozobot also has another potential; the robot uses Colour Codes to move in space and has its languages codified by colour. So, we proposed to students to codify Ozobot Evo by the use of Color Codes (age 9-12) and by coding it with the OzoBlockly platform (age 13-16) and letting Ozobot move in a space maze makes it reach as many exoplanets as possible, without falling into a black hole (Figure 3).

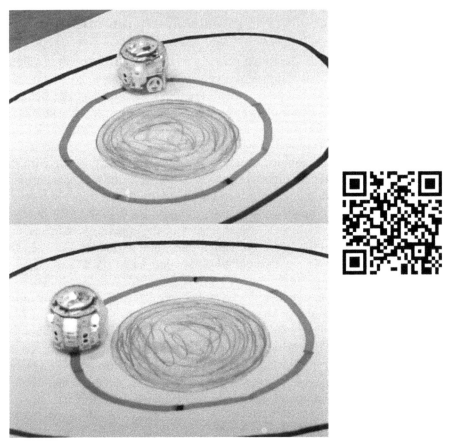

Figure 2: Simulation of the radial velocity method. When the star (Ozobot) approaches us, a blue signal will be recorded, when it goes away, we will record a red signal. Credits: L. Leonardi - INAF

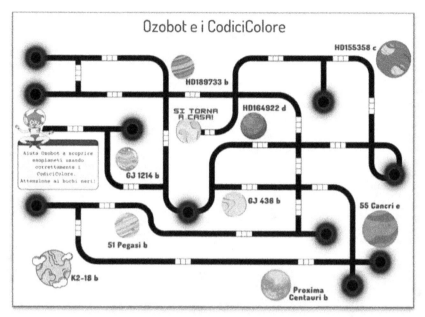

Figure 3: The maze of exoplanets. Credits: Discovery Education /L. Leonardi

Figure 4: The language of Ozobot

This maze can be solved in two ways: by using colour codes, or by programming with Ozoblockly. For our activities, we needed to codify the movements activated by colours (Figure 4): go right (blue-red- green), go left (green-black-red), go straight (blue-black-red), and win/exit (green-blue). Students could decide how to explore the maze to collect as many planets as possible.

As already mentioned, students can also challenge each other and solve the maze using the OzoBlockly platform (Figure 5).

See further details in the video by scanning the QR code in Figure 5.

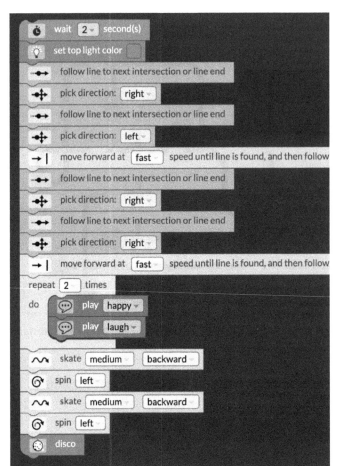

Figure 5: The code from the OzoBlockly platform and the qr code showing the resolution of the space maze with Colour Codes and OzoBlockly. Credits: L. Leonardi – INAF

Once all the blocks are in place, we move to the final part of the game. To celebrate Ozobot's return to planet Earth, as the activity helps the students also to understand the importance of our planet and the need to protect it, we reproduced Michael Jackson's famous Moonwalk dance move with the code "skate medium backward" and "spin left".

These two activities were developed in co-design with students during an INAF (National Institute for Astrophysics) series of activities for children/teenagers called "Astrokids" in front of about 30 students, and presented at the Europe Code Week 2020 attended by 8 schools from all over Italy.

We found that the use of ICT devices, like robotics, in astronomy education is effective in engaging teenagers and promoting autonomous learning processes. Furthermore, it develops students' critical thinking and digital skills, useful both for their personal lives and for their future careers. Moreover, ICT and educational robotics, in particular, allow them to develop soft skills such as problem-solving, self-awareness, and creativity.

3. Evaluation methodologies

When these activities were proposed for the first time, we didn't use any form of evaluation to gather observations and data. Our original aim was just to engage the students with science through innovative pedagogical activities. So, the only means we had to test the efficiency of the method was to observe how students interacted with the science of exoplanets - played by the little robots - and how they explained it to people visiting the science festival, where the activities with the Ozobot have been previewed in 2019.

After the activities, students were able to:
1.   Explain what is an exoplanet and how to find them
2.   Show how the science behind their observation works
3.   Reproduce the methods of exoplanetary research in a simple and engaging way
4.   Develop soft skills such as problem-solving (as already mentioned) and peer collaboration

However, after feedback received at the International Conference on Child-Robot Interaction 2023, in Milan, we decided to investigate how students approach science and new technologies such as educational robotics by means of these activities. So, we proposed the activity during the school project "Astronomia a Scuola" (trad. Astronomy at School) in February 2024 in Palermo (Italy), complemented by an evaluation plan, administering students' questionnaires to fill out before and after the activity with Ozobot.

In the responses to the post-questionnaire, 90,6% of the students indicate they have learnt new things about astronomy during the laboratory and 78,1% show that they have practised their programming skills (Figure 6). Moreover, the pre and post questionnaire asked students to write three words that come to mind when they think about "educational robotics". The most frequently words used about it, before the activity were: "future", "technology", "digital", "innovation", "simulation", "evolution", and "game".

Interestingly, there were some notable changes after the laboratory when the most frequently words used were: "opportunity", "discovery", "future", "technology", "digital", "innovation", and "simulation" (Figure 7).

Students realized the opportunity of this new educational application to explain the science that helps researchers make new discoveries.

During the laboratory, did you learn new tl astronomy?

During the laboratory, did you learn how t robot?

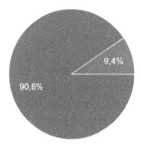

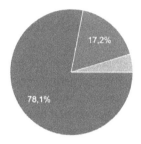

Figure 6: Results from the pre- and post-questionnaire

Figure 7: Two word clouds showing the words written by the students before (left) and after (right) the activities with Ozobot. The larger the word, the greater the number of students who wrote that same word

4. Conclusions

With the "digital transformation" of schools, digitalization acquires a wider meaning, aiming to optimize the logic of learning with more connectivity, devices and digital resources. Using a new "toolbox", however, is not enough. Digitalization has an "alphabet" in which computational thinking, a new syntax between logical and creative thinking, form the language we speak with increasing frequency (Sgambelluri, 2023).

The goal of our work is to support the school institution in fulfilling its mission, strengthening students' skills and learning experience by providing quality educational content and new tools created to facilitate schools, teachers and teaching in the new digital world.

Acknowledgements

The first author would like to express her gratitude to Serena Benatti, researcher at INAF in the field of exoplanetary research, for her scientific support in the development of these activities.

References

Bers, M. U., González-González, C., & Armaz-Torres, M. B. (2019). Coding as a playground: Promoting positive learning experiences in childhood classrooms. Computers & Education, 138, 130-145.

Chang, C. W., Lee, J. H., Chao, P. Y., Wang, C. Y., & Chen, G. D. (2010). Exploring the possibility of using humanoid robots as instructional tools for teaching a second language in primary school. Journal of Educational Technology & Society, 13(2), 13-24.

Gremigni, E. (2019). Competenze digitali e Media Education: potenzialità e limiti del piano Nazionale Scuola Digitale. Rivista Trimestrale di Scienza dell'Amministrazione, 1, 1-21

Kelly, W., McGrath, B., & Hubbard, D. (2022). Starting from 'scratch': Building young people's digital skills through a coding club collaboration with rural public libraries. Journal of Librarianship and Information Science, 09610006221090953.

Madsen, C., & West, R. M. (2003). Public communication of astronomy. Astronomy communication, 3-18.

Mayor, M., & Queloz, D. (1995). A Jupiter-mass companion to a solar-type star. nature, 378(6555), 355-359.

Montague, E., Xu, J., & Chiou, E. (2014). Shared experiences of technology and trust: An experimental study of physiological compliance between active and passive users in technology-mediated collaborative encounters. IEEE Transactions on Human-Machine Systems, 44(5), 614-624.

Sgambelluri, R. (2023). Competenze digitali e processi inclusivi per lo sviluppo di intelligenze collettive nella Scuola 4.0. PEDAGOGIA OGGI, 21(2), 107-116.

Wing, J. M. (2006). Computational thinking. Communications of the ACM, 49(3), 33-35.

United Nations (2015). Transforming Our World: The 2030 Agenda for Sustainable Development. Resolution Adopted by the General Assembly on 25 September 2015, 42809, 1-13.

QUI-BOT-H2O: Child-Robot Interactions Linking Chemistry And Robotics

Marta I. Tarrés-Puertas[1], Montserrat Pedreira Álvarez[2], Gabriel Lemkow-Tovias[3] and Antonio D. Dorado[4]

[1] Department of Mining, Industrial and ICT Engineering, Universitat Politècnica de Catalunya—BarcelonaTech (UPC), Manresa, 08242, Spain, 0000-0002-4473-6947, marta.isabel.tarres@upc.edu
[2] Research Group in Construction of Knowledge (GRECC), Faculty of Social Sciences, UVic-UCC, Manresa, 08241, Spain, 0000-0003-3680-4660, mpedreira@umanresa.cat
[3] Research Group in Construction of Knowledge (GRECC), Faculty of Social Sciences, UVic-UCC, Manresa, 08241, Spain, 0000-0003-3026-9919, GLemkow@umanresa.cat
[4] Department of Mining, Industrial and ICT Engineering, Universitat Politècnica de Catalunya—BarcelonaTech (UPC), Manresa, 08242, Spain, 0000-0003-0238-5867, toni.dorado@upc.edu

Abstract

The Qui-Bot H2O project aims to develop accessible, open-source robots and associated activities to cultivate early interest and engagement in scientific and technical fields by integrating chemistry and robotics. This initiative involves collaboration with experts in early childhood education, enabling the active participation of 70 young participants in exploring two innovative Qui-Bot H2O robotic prototypes: the non-humanoid Lab Qui-Bot constructed with LEGO elements, and the humanoid 3D Qui-Bot crafted without LEGO components. Both prototypes feature Human-Robot Interaction (HRI) components, including verbal color interpretations derived from chemical reactions and simulated facial expression changes through lighting. Built upon a maker philosophy and utilizing custom-made free software, the robots ensure accessibility for smaller institutions and simplify analysis for educators utilizing open-source tools. The design, implementation, and deployment of Qui-Bot H2O robots underscore affordability, ease of use, and gender-neutral design, specifically tailored for educational settings.

Keywords

Educational Robotics; STEM; Gender Stereotypes; Diversity; Digital Divide; Glass Ceiling; Feminist Robotics

1. Introduction

Statistics from the European Union (European Comission, 2022; MICINN, 2021) show that women's representation in STEM subjects at the university level hovers around 30% in the EU, dropping even lower to 15-20% in ICT (Information and Communication Technologies). Early education increasingly adopts robotics (Johal, Peng & Mi, 2020), such as Ozobot, programmable via colored pens for 4-year-olds, and Dash and Dot, programmable by children aged 5 and above.

The Qui-Bot H2O project aspires to boost an interest in STEM among children, drawing inspiration from the successful international program at Stanford University (Gerber et al. 2017). Qui-Bot H2O represents a pioneering effort to captivate young minds and instill a passion for STEM disciplines by merging chemistry, programming and robotics. The project introduces two novel robots, Lab Qui-Bot and 3D Qui-Bot, developed entirely from the ground up. Complementary chemistry-based activities and tailored programming challenges align with the early childhood education curriculum in Spain. The central goal of this initiative is to proactively introduce young children to programming and chemistry, mitigating the impact of societal stereotypes on their career choices, supported by scholarly research (Bian, Leslie & Cimpian, 2017).

No previous research has explored the intersection of educational robotics and chemistry, nor has it compared humanoid and non-humanoid chemical robot prototypes in early childhood contexts. Research in robotics within early childhood education, cited in works (Relkin, Ruiter & Bers, 2020; Yang, Ng & Gao, 2022), underscores the critical need to expand robot programming and computational thinking education across diverse educational settings. This research strongly advocates for integrated technology curricula seamlessly integrating these pivotal skills into educational frameworks. This integration seeks to infuse robot programming and computational thinking skills into educational activities for young learners. Moreover, it emphasizes the importance of educators in early childhood settings acquiring specialized training in robot programming, improving their traditional teaching skills. Equipping educators with this expertise enhances their ability to effectively guide and support children's learning experiences, fostering computational thinking among young learners.

2. Methodology

Our collaborative research within the Qui-Bot H2O project involves specialized universities spanning technology, science, pedagogy, economics and social administration. The research team comprises 15 institutions, with a pool of expertise of 22 dedicated members. In our preceding study (Tarrés-Puertas et al, 2022), we detailed the design process of Lab Qui-Bot, employing Lego-based components, and conducted comprehensive testing across a broad age spectrum from 3 to 18 years old.

In our latest research (Tarrés-Puertas, 2023), we introduce a groundbreaking prototype, 3D Qui-Bot—a humanoid robot programmed using wooden blocks. Additionally, we conduct supplementary tests to further assess its capabilities and functionalities. The study presented here primarily focuses on the outcomes derived from programming the 3D Qui-Bot, specifically addressing its significance in early childhood development. The research stages encompass various dimensions, including design intricacies, sensor integration, human-robot interaction (HRI) capabilities, technical programming guidance, and the implementation of inclusive interfaces. The design of all activities and materials has been meticulously developed, incorporating gender considerations as defined by subject matter experts.

All hardware instructions and materials are easily reproducible through the accessible open-access repository provided in UPC (n.d.). The training provided to 10 early childhood education teachers facilitated their use of the Lab Qui-Bot and the 3D Qui-Bot. These educators assumed the role of external evaluators, meticulously assessing behaviors and fostering children's creativity through storytelling, using stories, scenarios or narratives to enhance educational experiences involving robots. The scientific method and chemistry learning are acquired through guidelines and via small challenge-based games. The ability to modify the experiment and to command the robot through trial and error helps in developing their capacity for thinking, experimenting and working. The robot's design encourages them to think like a professional and improve their problem-solving abilities, promoting creativity in their resolution.

The activities of the 74 participating children were meticulously observed and captured via video for subsequent analysis by experts at the Lab0_6: Discovery, Research and Documentation Center for Science Education in Early Childhood (Lab 0_6, n.d.). One complete experience can be viewed at FibracatTV (2022). Furthermore, post-activity oral inquiries were employed to gauge varying levels of satisfaction. A sample of the Likert scale-based questions is available https://t.ly/4uIq (accessed April 2024). The gender equality has been ensured in the equal number of boys and girls who have participated in the testing. The initial study with Lab Qui-Bot involved boys and girls aged 3 to 6, with a total of 10 participants (4 boys and 6 girls). The sessions were conducted in groups guided by an educator, each lasting approximately 45 minutes. Activities included mixing liquids to observe color changes, the reverse process, and deriving colors from others. Results and evaluations are detailed in our prior work (Tarrés-Puertas et al, 2022), where it was noted that all children expressed 100% enjoyment and willingness to repeat the programming activities. The second study involved boys and girls aged between 3 and 6 years, with a total of 64 children participating, evenly split between 32 boys and 32 girls. The sessions were conducted in groups over eight sessions, each lasting approximately 60 minutes and guided by an educator. Children's actions were observed and recorded using a video camera for later analysis. Oral questions were used to evaluate their satisfaction during and after the activity.

3. Results

The Lab Qui-Bot and 3D Qui-Bot prototypes come with a range of sensors and actuators designed to enhance engagement and usability, as demonstrated in the technical prototype depictions provided in Figures 1 and 2. Supplementary materials, encompassing robot assembly instructions, a web interface, and a control app, have been developed utilizing open-source software and cost-effective consumables. While the Lab Qui-Bot is navigated via web/app interfaces, the 3D Qui-Bot employs colored wooden blocks for tangible programming.

The Lab Qui-Bot study findings illuminated a high level of enjoyment among children, coupled with a strong inclination to engage in programming activities anew. While some children initially encountered difficulties understanding computer commands and utilizing the mouse and touch sensor, they swiftly navigated these obstacles once comprehended. Notably, both boys and girls displayed keen interest in hands-on activities and exhibited collaborative behavior during group tasks. To refine the child experience, the interface underwent a redesign to align with user-friendly, inclusive, and captivating design elements tailored for children. These alterations included instances of robots executing mixing operations tied to real-world outcomes. In the later phase of our testing, we engaged with the 3D Qui-Bot. Children rapidly assimilated the programming codes (represented by colored cubes) and successfully crafted intricate action sequences, which they could modify as desired through manipulation of the color cubes. However, children under the age of 5 encountered hurdles in comprehending and anticipating complex chains of actions. Moreover, our findings underscored the pivotal significance of designing interfaces that are both user-friendly and inclusive, significantly enhancing children's engagement and learning experiences with technology.

Our approach acknowledges the power of storytelling in capturing human interest and fostering emotional connections with the robots. The study found that children preferred using the 3D humanoid Qui-Bot over the Lab/Lab2 QuiBots due to its intuitive interface and physical interaction with colored cubes. Younger children, under 5 years old, struggled with complex sequences but benefited from the more active and engaging nature of the 3D Qui-Bot activities. Boys tended to be more involved in decision-making during the activities, but both boys and girls enjoyed the games equally. Despite the success of the activities, improvements are needed in the design of the 3D robot, particularly in color scanning accuracy, mobility, durability, and robustness. Enhancements such as providing visual aids for programming and creating more challenging tasks using carpet "clippings" could further develop children's anticipatory and planning skills.

The promising initial outcomes of the Qui-Bot H2O project make significant contributions to several Sustainable Development Goals (SDGs):

- SDG 4: Quality Education - Through its emphasis on promoting computational thinking via robotics and interactive learning experiences, Qui-Bot H2O enhances the quality of education by providing teachers with innovative tools to teach scientific concepts effectively. By being designed for use in educational settings, it ensures that all students have access to inclusive and equitable education.
- SDG 5: Gender Equality - Qui-Bot H2O plays a crucial role in empowering women and girls in technology from an early age, thereby breaking down social barriers and stereotypes that hinder their development and progress in STEM fields.
- SDG 9: Industry, Innovation, and Infrastructure - By integrating real-world applications of robotics and technology into educational activities, Qui-Bot H2O promotes inclusive and sustainable industrialization. It highlights the importance of innovation and infrastructure development in driving economic growth and societal progress.
- SDG 10: Reduced Inequalities - Qui-Bot H2O aims to reduce inequalities by providing children from diverse social backgrounds with equal access to robotics education. By addressing social gender stereotypes associated with robotics, it ensures that all students have the opportunity to develop essential skills regardless of their gender. Additionally, Qui-Bot H2O aligns with the goals of the International Decade of Science for Sustainable Development 2024-2033 by fostering collaboration, providing equal opportunities, and promoting innovation in education, particularly for girls.

All the material created during the project's execution has been disseminated through various channels and made available to the general public via the repository published on the project's website (UPC, n.d.), aiming to promote scientific education in both formal and informal settings. As an openly accessible resource, it fosters educational innovation through collaborative efforts among institutions, providing accessibility to educators in chemistry, technology and all other areas, as well as students and researchers specializing in educational methodologies. Moreover, it enables seamless interaction, constructive feedback loops, and comprehensive evaluation among participants and instructors, thereby aiming to achieve a continuous improvement process.

Figure 1: Lab Qui-Bot   **Figure 2:** 3D Qui-Bot

4. Conclusions

The validation study, conducted by experts from Lab 0_6 as explained in the Methodology section, revealed that children exhibited a preference for the 3D humanoid Qui-Bot over the Lab Qui-Bot, perceiving it as more than a mere machine. Their heightened interest lay in robots capable of conducting chemical experiments as opposed to those performing solely mechanical functions. The children demonstrated an understanding of the robots' utility in handling hazardous liquids and ensuring precision. Additionally, the 3D Qui-Bot significantly aided the children in comprehending robot operations and facilitated logical arrangement using colored cubes. The validation also confirmed the efficacy of Human-Robot Interaction (HRI) features, such as verbalizing color outcomes.

Educators observed a variety of actions among children, notably related to executive functions like anticipation, self-correction, and working memory. The children actively engaged in hypothesis formulation, comparing their hypotheses with the robot's actions, and occasionally imitated the robot's movements to anticipate or reproduce their moves. Collaborative endeavors were prevalent, with children offering each other advice on effectively programming the robot.

Researchers are actively exploring the integration of advanced HRI capabilities into Qui-Bots to enable novel interactions. Their future agenda includes the introduction of Qui-Bots into educational settings, teacher training initiatives, and the formulation of an ethical code based on existing literature (Smakman, Vogt & Konijn, 2021).

The research unveiled limitations in utilizing Lego Mindstorms kits for robotics tasks, reported by educators. Challenges encompassed the necessity to acquire multiple kits for simultaneous group use, teachers' hesitation in adopting custom interfaces despite prior Lego software experience, and the prerequisite for 3D design proficiency to replicate the 3D Humanoid and procure materials. To overcome these limitations, the research advocates for interactive training sessions lasting at least two hours, comprising hands-on assembly and rigorous testing of robots from inception, thereby enhancing the project's applicability and effectiveness. Future endeavors could explore the possibility of children constructing the Lab Qui-Bot from scratch during early childhood, fostering a tangible learning experience. Additionally, partial 3D printing of robot components, particularly for the humanoid model, warrants further investigation and development. Educational outreach materials are under development to encourage early engagement with robots among children and maximize their benefits, as depicted in Figure 3. The creation of logos, brochures and promotional materials has been crafted by illustrators specialized in engaging children's audiences.

Future work is committed to pursuing additional objectives that advance feminist robotics and promote inclusivity and equity in the technology sector as follows. Exploring Robotics from a Feminist Perspective: We will conduct research and develop educational programs inspired by feminist principles outlined in "Data Feminism" (D'Ignazio & Klein, 2020). By challenging sexism and promoting inclusion in robotics, we strive to empower every girl to make a significant impact in technology.

Expanding Data Sampling and Deployment of Qui-Bots: We will broaden our reach by deploying Qui-Bots to more educational centers and the general public. This expansion will allow us to gather diverse perspectives and feedback, enhancing the effectiveness of our robotics initiatives.

Feminization of the Digital Sector: Through educational policies, we aim to feminize the digital sector, starting from school level. By empowering girls in STEAM fields, we seek to create a more inclusive and diverse workforce in the digital industry.

Breaking Down Stereotypes: Our efforts will continue to challenge stereotypes associated with technological studies. By promoting inclusivity and challenging traditional gender roles, we aim to create an equitable environment for all individuals interested in technology careers.

Prioritizing Equality: Equality remains a paramount value in our initiatives. We are dedicated to ensuring equal access to educational and professional opportunities in robotics, regardless of gender, economic status or other distinctions.

Counteracting Gender Division and Stereotypes: Building on the findings of the UNESCO report, (West, Kraut & Chew, 2019), we will actively work to counteract gender division and stereotypes. Our goal is to foster an inclusive environment where individuals of all genders feel valued and empowered in technology.

Improving HRI Aspects of Qui-Bots: We will enhance the Human-Robot Interaction (HRI) aspects of Qui-Bots to create feminist robots that effectively interact with diverse user groups. Drawing from experiences like those of the University of Sweden, (Winkle, n.d.), we will ensure inclusivity and accessibility for all users.

Figure 3: Education outreach material of Qui-Bot H$_2$O project

Acknowledgements

The Qui-Bot-H2O project is supported by the Spanish Ministerio de Economía y Competitividad under Grant FECYT2021-15626, Line of action 2. Education and scientific vocations.

References

Bian, L., Leslie, S.-J., & Cimpian, A. (2017). Gender stereotypes about intellectual ability emerge early and influence children's interests. Science, 355(6323), 389-391.

D'Ignazio, C., & Klein, L. F. (2020). Data Feminism. The MIT Press. Retrieved from https://mitpress.mit.edu/9780262044004/ (Accessed on April 24, 2024)

European Commission. (2022). Bridging the gender gap in STEM: Strengthening opportunities for women in research and innovation. Retrieved from http://bit.ly/408alx5 (Accessed April 24, 2024).

FibracatTV. (2022, October 4). Lab-06 Qui-Bot Robot [Video file]. Retrieved from https://www.youtube.com/watch?v=KRzUT0hVFE4 (Accessed on April 24, 2024)

Gerber, L. C., Calasanz-Kaiser, A., Hyman, L., Voitiuk, K., Patil, U., et al. (2017). Liquid-handling Lego robots and experiments for STEM education and research. PLOS Biology, 15(3), e2001413.

Johal, W., Peng, Y., & Mi, H. (2020). Swarm Robots in Education: A Review of Challenges and Opportunities. In Proceedings of the 8th International Conference on Human-Agent Interaction (HAI '20) (pp. 272–274). Association for Computing Machinery.

Lab_06: Discovery, Research and Documentation Center for Science Education in Early Childhood, located within the Faculty of Social Sciences in Manresa (UVic-UCC). (n.d.). Retrieved April 24, 2024, from https://lab06.umanresa.cat/

MICINN - Spanish Ministry of Science and Innovation. (2021). Scientific Report Women in Science. Retrieved from https://www.ciencia.gob.es/dam/jcr:dc8689c4-2c47-4aaf-97ce-874bd0b5a081/Cientificas_en_Cifras_2021.pdf (Accessed April 24, 2024).

UPC- OCW Qui-Bot H2O project Home Page. Universitat Politècnica de Catalunya · BarcelonaTech (UPC). (n.d.). https://quibot.upc.edu (Accessed April 24, 2024).

Relkin, E., de Ruiter, L., & Bers, M. U. (2020). TechCheck: Development and validation of an unplugged assessment of computational thinking in early childhood education. Journal of Science Education and Technology, 29, 482–498.

Smakman, M., Vogt, P., & Konijn, E. A. (2021). Moral considerations on social robots in Education: A multi-stakeholder perspective. Computers & Education, 174, 104317.

Tarrés-Puertas, M. I., Merino, J., Vives-Pons, J., Rossell, J. M., Pedreira Álvarez, M., Lemkow-Tovias, G., & Dorado, A. D. (2022). Sparking the Interest of Girls in Computer Science via Chemical Experimentation and Robotics: The Qui-Bot H2O Case Study. Sensors, 22(10), 3719.

Tarrés-Puertas, M.I., Costa, V., Pedreira Alvarez, M., Lemkow-Tovias, G., Rossell, J.M., Dorado, A.D. (2023). Child–Robot Interactions Using Educational Robots: An Ethical and Inclusive Perspective. Sensors, 23, 1675.

West, Mark, Kraut, Rebecca, Chew Han Ei, I'd blush if I could: closing gender divides in digital skills through education, UNESCO, EQUALS Skills Coalition, 2019,

Winkle, K. (n.d.). Can Feminist Robots Challenge Our Biases? KTH Royal Institute of Technology and Stockholm University. Retrieved from https://spectrum.ieee.org/human-robot-interaction (Accessed on April 24, 2024)

Yang, W., Ng, D. T. K., & Gao, H. (2022). Robot programming versus block play in early childhood education: Effects on computational thinking, sequencing ability, and self-regulation. British Journal of Educational Technology.

"On board a photon": an educational escape room based on robotics and unplugged coding to discover the journey of light across the Solar System

Claudia **Mignone**[1], Silvia **Galleti**[2], Laura **Leonardi**[3], Federico **Di Giacomo**[4], Maria Teresa **Fulco**[5] and Maura **Sandri**[6]

[1] *INAF – Istituto Nazionale di Astrofisica - Sede Centrale, viale del Parco Mellini 84. 00136, Roma, Italy, 0000-0002-5525-314X, claudia.mignone@inaf.it*
[2] *INAF – Istituto Nazionale di Astrofisica - Osservatorio di Astrofisica e Scienza dello Spazio, via Gobetti 93/3, 40129 Bologna, Italy, 0000-0003-3825-2273, silvia.galleti@inaf.it*
[3] *INAF – Istituto Nazionale di Astrofisica - Osservatorio Astronomico di Palermo Giuseppe S. Vaiana, Piazza del Parlamento, 1, 90134 Palermo, Italy, 0000-0001-6555-4934, laura.leonardi@inaf.it*
[4] *INAF – Istituto Nazionale di Astrofisica - Osservatorio di Astrofisica e Scienza dello Spazio, via Gobetti 93/3, 40129 Bologna, Italy, 0000-0002-9180-1019, federico.digiacomo@inaf.it*
[5] *INAF – Istituto Nazionale di Astrofisica - Osservatorio Astronomico di Capodimonte, Salita Moiariello, 16, 80131 Napoli, Italy, 0000-0001-8432-2952, mteresa.fulco@inaf.it*
[6] *INAF – Istituto Nazionale di Astrofisica - Osservatorio di Astrofisica e Scienza dello Spazio, via Gobetti 93/3, 40129 Bologna, Italy, 0000-0003-4806-5375, maura.sandri@inaf.it*

Abstract

This work presents an educational escape room designed by INAF, the Italian public research institute dedicated to the study of the Universe, as part of its third mission to engage society at large with astronomy and the collective scientific endeavour of making sense of the cosmos. The experience, aimed at secondary school students (aged 11-18), combines coding and robotics in a playful way with a scientific narrative – the journey of light in the Solar System – to discover new topics in astronomy while practising computational thinking. The leitmotiv of the activity is the journey of a particle of light, or photon, from its generation in the core of the Sun and its complex trek across the solar interior, all the way across space to one of the Solar System's planets that reflect it towards Earth, where it eventually lands on a telescope that produces an image of that particular planet. Each challenge is centred on a key step in this journey, using small robots, decoding clues from encrypted devices and solving unplugged coding exercises. We present the design and implementation of the escape room as well as results from a preliminary evaluation of the project conducted at the Genova Science Festival, where it premiered in 2022.

Keywords

Astronomy Education, Public Engagement, Escape Room, Unplugged Coding, Educational Robotics, Storytelling

1. Introduction

Escape rooms are environments, either virtual or real, in which participants have to seek out clues to solve a series of challenges that eventually lead them to get away from that environment. Originally created as entertainment experiences, they are proving increasingly popular as educational tools since they require a combination of different skills and expertise, promote teamwork and, depending on the theme, may be used to activate several curricular skills as well as soft skills, such as problem solving, attention to details and interpersonal communication (Vizzari, 2022).

Computational thinking, defined as a practice that "involves solving problems, designing systems, and understanding human behaviour, by drawing on the concepts fundamental to computer science'' (Wing, 2006), is an essential ability to make sense of the world, interact with and co-create technology. As demonstrated by extensive research in this field, coding and educational robotics are suitable tools to create practical and motivating activities that can fuel interest and curiosity (Eguchi, 2010) in students of all orders and grades. Besides, educational robotics is featured prominently in today's school curriculum as a tool to foster important soft skills such as cognitive and personal development and teamwork (Alimisis, 2013). In this context, the use of small, developmentally appropriate robots is encouraged to practise sequencing ability (Kazakoff, Sullivan & Bers, 2013), debugging and problem-solving skills (Bers, Flannery, Kazakoff, & Sullivan, 2014), along with unplugged coding activities that can be introduced in the classroom well before computer programming (Lee & Junoh, 2019), exploiting the strong link with the narrative approach that these educational tools share (Rusk, Resnick, Berg, & Pezalla-Granlund, 2008). With more advanced students, the same tools can be used to support in-depth activities exploring scientific topics, e.g. physics concepts in secondary school (Souza & Duarte, 2015).

The use of games in education (game-based learning) motivates learners, engages them in a wide range of ways, can be adapted to a variety of different contexts and encourages taking risks and trying new things in a shielded environment where failure is present by design so it can be addressed in a graceful manner (Plass, Homer, & Kinzer, 2015). Likewise, storytelling can help make a subject more meaningful, relevant and accessible: narrative thinking is based on the logic of human actions, aiming at understanding the human condition, and therefore reflects the experience of problem solving. Stories support learners in the organisation of knowledge, the construction of meaning and the resolution of problems, which is especially important in the context of science education (Avraamidou & Osborne, 2009). Education research has demonstrated that narrative-based resources are effective for teaching complex and intricate topics (Gulay, 2010; Mangione, Capuano, Orciuoli, & Ritrovato, 2013). Digital storytelling combines the power of narrative with images, video, digital devices and web-based technologies, enabling learners to organise, express, re-mediate, and share ideas and knowledge in a creative, original way (Petrucco & De Rossi, 2009). The overlap of digital storytelling and game-based learning can be explored in the classroom (Alexander, 2011) to practise media literacy and awareness, along with a set of social skills and cultural competences that are developed through collaboration, such as: play, performance, simulation, appropriation, multitasking, distributed cognition, collective intelligence, judgement, transmedia navigation, networking and negotiation (Jenkins, 2009). These skills and competences arise as critical also from research in educational neuroscience (Rivoltella, 2012), calling for the adoption of media- and technology-rich pedagogies that involve games and multiple codification systems (images, sound, tactile, spatial) along with teamwork, embodied and peer learning, embracing digital tools as an opportunity to rethink the educational practice for the 21st century (Maragliano & Pireddu, 2015).

The escape room described in this work, titled "On board a photon" (in Italian: "A cavallo di un fotone"), combines unplugged coding and robotics in a playful way with a scientific narrative – the journey of light in the Solar System – to approach new astronomical concepts while practising computational thinking as well as enabling students to solve problems and work in teams. Aimed primarily at secondary school students (aged 11-18), this escape room is an interactive serious game experience to be played in a physical environment, using a number of analogue objects along with educational robotics and unplugged coding exercises.

2. The escape room design and structure

This escape room was designed by the Italian National Institute for Astrophysics (INAF) Play Coding working group in response to the open call launched by the Genova science festival[4] in 2022 around the theme "Languages", based on extensive experience with coding and robotics in astronomy education (Sandri et al., 2023). The leitmotiv of the activity is the

---

[4] The Genova Science Festival is one of the main public engagement initiatives in Italy: https://www.festivalscienza.it/

journey of a particle of light, or photon, from its generation in the core of the Sun and its complex trek across the solar interior, all the way across space to one of the Solar System's planets that reflects it towards Earth, where it eventually lands on a telescope that produces an image of that particular planet. In the classical storytelling model of the Hero's Journey (Campbell, 1949), the escape room, which takes place in a real space, is structured around a series of challenges, each centred on a key step in a photon's journey towards an astronomical image: first the Sun, then the planets and, finally, a telescope on Earth.

The main character of the narrative is an inanimate object: an elementary particle, in this case the photon, following an approach that has been used on occasions in physics education and outreach resources (e.g. Peduto, 2012). Participants are sorted into teams, and each team follows the journey of a different photon. According to the storytelling design model described by Mangione, Orciuoli, Pierri, Ritrovato, & Rosciano (2011), the various steps of the activity can be mapped to the story structure as follows: in the *beginning*, hydrogen is packed at the core of the Sun, where density and temperatures are so high that nuclear fusion reactions turn hydrogen into helium, producing photons as a result. This is the ordinary routine that is interrupted by the *call to adventure*: photons must set on an exciting journey across the Solar System and reach telescopes on planet Earth. However, the journey is ripe with *problems and difficult tasks to solve*: first, photons must get out of the Sun, then they must find the route that takes them to a given planet, then get back to Earth and finally find the right telescope to land upon. All stories are marked by one key *transformation* of the main character, who has to undergo an important change in order to fulfil its goal: in this case, after having visited a planet, photons must be reflected towards Earth and reach a particular telescope there. Finally, upon reaching their telescope destination, photons are captured by photographic instruments and produce a photo of the planet they have visited along the journey, creating *closure* to internalise the adventure each team has experienced throughout the activity.

The main goals of the activity are: i) learning new astronomical concepts like distance scales in the Solar System, the time taken by light to cross such distances and how telescope observations are performed; ii) practicing computational thinking; iii) activating soft skills such as problem solving and group collaboration. Therefore, the various steps in the scientific narrative are combined with coding and robotics challenges in a playful way, to encourage participants to discover new topics while at the same time developing computational thinking. Each challenge includes classic analogue objects used in escape rooms, such as locked devices, treasure chests, puzzles and maps, along with activities based on educational robotics (Eguchi, 2010) and unplugged coding, i.e. a computer science activity that does not include actual programming and can be done offline without a computer (Bell, Alexander,

Freeman, & Grimley, 2009). In line with the festival theme, clues are provided in the form of programming language.

2.1. First challenge: get out of the Sun

The first challenge in this escape room consists of getting out of the Sun. After players are sorted into teams, they are invited to watch a short animated video about the nuclear fusion reactions taking place at the core of the Sun and how photons, the particles of light that are produced as a result of such reactions, travel across the solar interior. The journey of photons from the centre of the Sun to its surface may take hundreds of thousands of years, first through the radiative zone, where they experience frequent collisions within the dense solar plasma that change their direction in random ways, resulting in a zig-zag trajectory, and later through the convective zone, where photons are captured by large-scale flows that eventually take them to the solar surface (Carroll & Ostlie, 1996).

To reproduce this lengthy journey in the escape room, the Sun is rendered as a 3-metre across custom-design carpet featuring a labyrinth (Figure 1). In order to overcome the first challenge, students must correctly program a Bee-Bot or Blue-Bot – an educational robot that executes simple sequences of commands such as forward, left and right, designed specifically to be used by young children (Pekarova, 2008). The instructions to program the robot are provided in coding language as a "universal move" card with the following text: "if there is space ahead, move forward; else, if there is space on the left/right, turn left/right; else, turn right/left". We chose to employ a simple educational robot, aimed at much younger children than the students involved in this activity, to enable them to focus on the coding logic behind the instructions and how that relates to the complex journey of photons inside the Sun, rather than on the specific robot commands. Different teams take on this challenge at the same time, however, each team must follow a different path on the labyrinth, according to the team's assigned colour. If they program their photon/robot to exit the labyrinth correctly, it will move along a series of boxes containing letters. These letters reveal a password that will be needed to solve the next puzzle: opening an encoded locking device called cryptex (seen at the edge of the Sun in Figure 1).

2.2. Second challenge: find the planet

Inside the cryptex, each team finds another code that can be solved through the unplugged coding practice of pixel art (Capecchi, Gena, & Lombardi, 2022). Pixel art is defined as "any drawing that emphasises its underlying pixel structure" and it can be represented as a run-length encoded sequence to indicate the colour of each pixel and specify how many pixels should be painted in each colour (Bogliolo, 2017). In this case, the pixel art codes represent different planets in the Solar System[5], which students can solve using squared sheets and coloured pencils, in order to find their next destination. After this challenge, each team is sent to a different planet (Figure 2).

Figure 1: The custom-made carpet with the labyrinth to get out of the Sun by correctly programming a Blue-Bot. On the edge of the Sun, eight cryptex devices are visible. Credits: INAF

2.3. Third challenge: find the telescope on Earth

Next to each planet, students find a locked treasure chest, which can be unlocked by solving quizzes about that particular planet; all clues to solve the quizzes can be found in the room, on roll-up installations. Inside the box are various objects: two torches (one working with visible light, one with ultraviolet light), pieces of a puzzle, and a compact disc – some of the objects are merely distractors. If combined appropriately, the objects will reveal yet another code: a set of terrestrial coordinates. With the help of a map, the coordinates allow players to locate a telescope on Earth, which is the final destination of each photon's journey.

---

[5] The pixel art codes for Solar System planets, as well as other astronomical images, are available are available on the INAF Play website: https://play.inaf.it/tag/pixel-art/

2.4. Fourth challenge: reach the right telescope

In the final challenge, players must program another small robot, an Ozobot Evo (Balaton, Cavadas, Carvalho, & Lima, 2021), to move across a specially designed maze and reach the correct destination among eight different telescopes (Figure 2). The Ozobot Evo is one of the smallest programmable robots: it can be instructed to follow a black line, and students can guide it towards different directions by applying appropriate colour codes to the empty spaces placed along the maze. When participants complete the last challenge, reaching the correct telescope on Earth with the Ozobot, the journey of their photon is also complete, and they receive a key that symbolically marks the exit from the escape room. The key also opens a box that contains a postcard (prize) featuring the image of the planet each team has "visited" along their journey, as actually observed by the telescope they have reached. Thus, the journey of the photon, born inside the Sun and reflected by the planet towards a telescope on Earth, ends in a beautiful image that summarises its adventure.

2.5. Time span and target groups

The game was designed to last a maximum of one hour, with different teams of 4-5 players each playing at the same time and competing to exit the escape room. A stopwatch at one end of the room would remind players of the time left to complete the challenges. The game design was aimed primarily at students in the lower secondary (grade 6th to 8th) and upper secondary (grade 9th to 13th) echelons of the Italian school system. Besides, following guidelines from the festival, it was designed so that the challenges and tools could also be accessible to late primary school students (aged 10 or older). Balancing the experience so that it would be educational and entertaining for such a wide variety of age groups is by no means trivial, so we assumed different teams would move through the game at a different pace, depending on their age and skills. Indeed, many teams reached the end of the escape room well ahead of time, within 45 minutes or less.

Figure 2: In the foreground, one of the pixel art sheets, representing the planet Saturn (only half of the drawing has been completed). In the background, students are programming the Ozobot maze. Credits: INAF

3. Evaluation during a science festival

The escape room, which was the INAF Play Coding team's first experiment with this type of activity, premiered for 13 days at the Genova Science Festival in October 2022. There, the activity was facilitated by a team of festival assistants, after being trained by the team who had designed the escape room. Almost 2000 participants took part in this activity during the festival, including around 900 students from different education levels (4 groups from primary school; 26 from lower secondary school; 11 from upper secondary school) and different Italian regions (Liguria, Piemonte, Lombardia and Campania), as well as one group of students from Sweden (with support from festival facilitators for the English translation).

Students were asked to fill in a pre- and post-activity anonymous questionnaire to assess whether the activity reached its main goals. Around 300 students filled in the pre-activity questionnaire and around 100 students (about 10% of all students who participated in the escape room) filled in the post-activity questionnaire. Mainly students from lower secondary school (61,8%) and upper secondary school (28,4%) filled in the post-activity questionnaire, with a slight prevalence of female students (49,5%) with respect to male students (42,6%). In addition to that, 260 anonymous sticky notes with comments were left in a dedicated box.

3.1. Qualitative analysis of questionnaires

In response to the post-activity questionnaire, 84% of the sample indicated that they learnt new things about astronomy during the escape room and 80% that they practised their computational skills (Figure 3). Besides, 80% of respondents declared that their experience would be best described as "having fun", 68% as "learning new things about the Universe", 60% as "playing a game in a group", and 22% as "learning how to solve complex challenges" (it was possible to indicate more than one option).

For most participants who completed the post-activity questionnaire, it was the first time they took part in a science themed escape room, and the majority indicated that they would like to do that again in the future. However, a couple of answers in the free comment field indicated that a different kind of escape room was expected, pointing out that there was no actual "exit door" and that the activity rather resembled a treasure hunt. Overall, the experience was widely praised and positively appreciated: most frequently used words in the free comment box were "nice" (23%), "fun" (20%), "interesting" (10%) and "learning" (10%). When asked what they enjoyed the most, the most frequent answers were: "robots" (22%), "sun" (15%), "getting out" (14%), "coding" (12%), "solving" (5%) and "challenge" (5%); when asked what they enjoyed the least, the most frequent answers were: "nothing" (18%), "planet" (14%), "colouring" (12%), "finding" (12%), "sun" (5%) and "getting out" (5%).

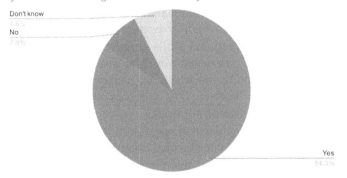

During your visit to the escape room "On board a photon", did you learn new things about astronomy?

Don't know
7.8%

No
7.8%

Yes
84.3%

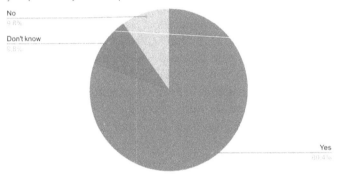

During your visit to the escape room "On board a photon", did you practise your computational skills?

No
9.8%

Don't know
9.8%

Yes
80.4%

Figure 3: Self-assessment of new astronomy concept learning (top) and computational skill practice (bottom) by the participants

Regarding the difficulty of the various challenges, participants were asked to rate them via a 5-point Likert scale from "very easy" to "very difficult" (Figure 4). The first challenge – getting out of the Sun using the Blue-Bot – was deemed either "too easy" or "easy" by 75% of respondents, with 25% finding it "neither easy nor difficult" and less than 10% "difficult". The second challenge – finding the planet destination with pixel art – was deemed either "too easy" or "easy" by 65% of respondents, with 20% finding it "neither easy nor difficult" and around 10% either "difficult" or "too difficult". The third challenge – finding the telescope destination on Earth via a quiz, various objects, coordinates and a map – was deemed "easy" by 40% of respondents, with another 40% evenly split between "too easy" and "neither easy nor difficult", and 10% finding it "difficult". Finally, the fourth challenge – reaching the telescope with Ozobot – has a more evenly split rating, with 30% finding it "too easy", 20% "easy", 25% "neither easy nor difficult" and another 20% "difficult". Besides, a handful of participants mentioned in the free comment field that they found some (or all) challenges too easy: these were either secondary school students or adults; on the other hand, one comment left on the sticky notes referred to the entire experience as "too complicated".

The questionnaire also asked participants to write three words that come to mind when they think about i) the Solar System, ii) the journey of light in the Universe, iii) coding and iv) robots, both before and after the activity. We performed a qualitative analysis of the answers, with the caveat that the sample who responded to the pre-activity questionnaire does not coincide with and is three times larger than the post-activity one (see Tables 1 and 2).

How do you evaluate the complexity of the various challenges?
0: don't know; 1: very easy; 2: easy; 3: neither easy nor difficult; 4: difficult; 5: very difficult

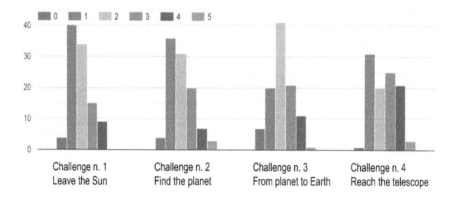

Figure 4: Perception of challenge complexity by escape room participants

The most frequently used words linked to the Solar System were "Sun", "planet", "Earth" and "star", with no substantial change between pre- and post-activity answers. The most frequently used words concerning the journey of light in the Universe were "speed / fast", "light", "space" and "time", with one notable difference between pre- and post-activity answers: the use of the word "photon/photons", which increased from being used in 5% to 18% of the answers. The most frequently used words when thinking about robots are "technology", "intelligence", "future" and "artificial" (in more than half of the cases, "intelligence" is used in association with "artificial"), with a non negligible increase in the occurrence of the words "programming" (from 12% to 17% of the responses) and "coding" (from 0.3% to 5%). Finally, some changes are observed in the most frequently used words related to coding between pre- and post-activity answers: while the most used word in both cases is "computer", it experiences a decrease from 36% to 25% of the answers, and the second most used word, "technology", also decreases from 16% to 9% of the answers; the words "programme / programming" remain stable around 13-14% of the answers, while the use of the word "robot" increases from 8% to 18% of the answers. It is also worth noting a few subtle differences between some of the less frequent words associated with coding, such as "fun" and "interesting", which increase from 2-3% to 5% of the answers, while the word "difficult" has a slight decrease from 3% to 2%.

Table 1
Most frequently mentioned words by escape room participants about the Solar System (left) and the journey of light in the Universe (right) before and after the activity, respectively

Words that come to mind when thinking about the Solar System		Words that come to mind when thinking about the journey of light in the Universe	
PRE	POST	PRE	POST
Sun (195/309)	Planet (57/103)	Speed / fast (205/309)	Speed / fast (73/103)
Planet (186/309)	Sun (52/103)	Light (67/309)	**Photon (19/103)**
Earth (76/309)	Earth (21/103)	Space (38/309)	Light (12/103)
Star (49/309)	Star (17/103)	Time (35/309)	Time (12/103)
Space (40/309)	Light (11/103)	Star (30/309)	Space (9/103)
Light (32/309)	Space (8/103)	[...] Photon (15/309)	Star (9/103)

Table 2
Most frequently mentioned words by escape room participants concerning robots (left) and coding (right) before and after the activity, respectively

Words that come to mind when thinking about robots		Words that come to mind when thinking about coding	
PRE	POST	PRE	POST
Technology (95/309)	Intelligence (28/103)	Computer (112/309)	Computer (26/103)
Intelligence (83/309)	Technology (26/103)	Technology (50/309)	**Robot (19/103)**
Future (52/309)	Future (20/103)	Programming (39/309)	Codes (14/103)
Artificial (49/309)	**Programming (17/103)**	Informatics (33/309)	Programming (14/103)
Metal (46/309)	Artificial (16/103)	Numbers (32/309)	Numbers (12/103)
Programming (36/309)	Mechanics (8/103)	Codes (27/309)	Technology (10/103)
Machine (25/309)	Machine (7/103)	**Robot/Robotics (24/309)**	Commands (6/103)
Mechanics (22103)	Computer (6/103)	Logic (17/309)	Informatics (6/103)
Innovation (15/309)	Innovation (6/103)	[...] Fun (8/309)	[...] Fun (5/103)
Informatics (13/309)	**Coding (5/103)**	[...] Difficult (8/309)	[...] Interesting (5/103)
[...] Coding (1/309)	Gears (5/4)	[...] Interesting (5/309)	[...] Difficult (2/103)

3.2. Qualitative analysis of sticky notes

The analysis of the 260 handwritten comments posted in an anonymous feedback box indicates an overwhelmingly positive sentiment (94%), with only a few notes containing negative (2%) or neutral (2%) sentiment, or off-topic comments (2%). Among the most mentioned words are "fun" (33%), "nice" (30%), "very nice" (17%) and "interesting" (13%). It is also worth noting a few among the less frequently mentioned themes, such as the importance of "teamwork" (appearing in 5% of the comments) – e.g.: "I really enjoyed the experience because it was nice to solve it as a team" – and the motivational role of "competition / winning" (3%) – e.g.: "Great experience (We won!!!)".

Among notable comments, one sticky note reads: "It was very exciting to play with the team, the challenges were great, but the biggest surprise was when we discovered the postcard and how the picture had been made" – which summarises the take-home message of the narrative underpinning the educational experience, emphasising the key role of the story's closure. Another comment emphasises the importance of the playful atmosphere and the relevance of game-based learning: "I had fun because by playing I can appreciate a somewhat difficult topic such as science". Finally, a handful of notes also mention "would repeat it" (8%), "would recommend it" (2%), "too easy" (3%) and "too difficult" (1%).

3.3. Discussion

The self-assessment completed by participants in the questionnaires, along with the free comments provided both in the questionnaire and the sticky notes, point towards an overall positive and enjoyable experience, indicating that the escape room reached its three main goals: i) learning new astronomical concepts; ii) practising computational thinking; iii) activating soft skills such as problem solving and group collaboration. The four challenges were perceived as being of increasing difficulty as the activity proceeded. Some participants deemed some (or all) of the challenges too easy, while at least one participant found the overall experience too complicated: these discrepancies are most likely an effect of the very broad target group of the activity, which during the festival was offered to students aged 10 or older, as well as adults.

A qualitative analysis of word clouds from three-word answers in response to four different topics encountered during the activity indicates that some changes might have happened as a result of the experience. For example, the increased occurrence of the word "photon" in relation to the journey of light in the Universe in the post-activity questionnaire is evidently an effect of having heard the word being used on several occasions during the activity, not least in the escape room's title itself. Besides, the increased occurrence of words such as "coding" and "programming" when thinking about robots, and of the word "robot" when thinking about coding, might point to students having seen the two concepts at play simultaneously during the activity. The mild increase in the occurrence of words such as "fun" and "interesting" in relation to coding, along with the mild decrease of the word "difficult", might be a result of the playful activity that presents coding as an entertaining topic; however, it might also be simply a statistical fluctuation.

In the future, an in-depth evaluation will also explore learning outcomes, asking pre- and post- questions about photons, their journey across the Solar System and how astronomical images of planets are taken, to assess whether there are any appreciable changes in the understanding of these concepts after having participated in the escape room. We also envision observations of how teams experience the game, measuring the time spent solving each challenge and cross-correlating it with the narrative elements, to validate the adopted storytelling design model in connection with coding and robotics.

4. Conclusions

This work provides an example of a learning activity aimed at secondary school students (aged 11-18) combining astrophysics content with unplugged coding and educational robotics, drawing upon game-based learning and storytelling. The activity, designed in the form of an escape room, explores topics such as the production of light and energy by the Sun, the propagation of light across space, planets in the Solar System, and astronomical telescopes on Earth, enabling students to practise digital skills and computational thinking at the same time.

Based on the premiere at the 2022 Science Festival in Genova and the qualitative analysis of data collected there with a representative sample of participants, the escape room reached its objective of providing an entertaining educational experience, engaging students with new astronomical concepts while practising computational thinking and digital skills in a fun activity.

The escape room can be replicated in other contexts: in the spirit of Open Science, the instructions to play, as well as all files necessary to reproduce the installation, have been made freely available online in Italian and English on the INAF Play website[6], along with advice on how to replicate a low-cost version, e.g. by printing on A4 sheets instead of high-quality roll-ups, or using unplugged coding with forward / left / right cards (Bogliolo, 2020) instead of the robots. A manual with teacher instructions will be made available soon (Mignone et al., in preparation). To further exploit the power of narrative in

---

[6] The escape room description and materials to replicate it are available on the INAF Play website: https://play.inaf.it/en/on-board-a-photon/

the learning of complex concepts, the activity can be followed by a collective exercise of reflection and re-mediation, during which each team re-tells the story of their photon using the media of their choice (text, video, performance, illustration, etc.) to consolidate the acquired knowledge and share it with peers while practising digital storytelling at the same time, alternating small-groups and classroom inputs (Di Blas, Paolini, & Sabiescu, 2010).

The activity was replicated in September 2023 by a team of INAF public outreach professionals during the Trieste Next Science Festival, with 4 secondary school classes participating, and during the European Researchers' Night, open to the general public as part of the European Commission funded Sharper project in Trieste. A more structured pre- and post-activity questionnaire has been prepared for this occasion to attempt a quantitative evaluation of the learning outcomes, however, only a dozen students have filled in the form so the data have not yet been analysed. Meanwhile, at least one school and one planetarium have already inquired about the possibility of replicating the escape room at their premises. Pending funding availability, the team is considering a possible adaptation of the game to a virtual (digital) environment to be used more widely in the classroom, which would overcome budget and spatial limitations often present in schools and also make it accessible to individuals with disabilities or who are homebound.

Acknowledgements

We are grateful to Festival della Scienza for accommodating this activity in their programme, to the festival animators who facilitated it, and to all participants who experienced it and provided useful feedback. We wish to thank the organisers of the International Conference on Child-Robot Interaction CRI23 for the opportunity to present this work and the two anonymous referees for carefully reading the manuscript and for the constructive suggestions that improved the scope of the paper.

References

Alexander, B. (2011). The New Digital Storytelling: Creating Narratives with New Media. Praeger, ABC-CLIO, Santa Barbara, CA

Alimisis, D. (2013). Educational robotics: Open questions and new challenges. Themes in Science & Technology Education, 6(1), pp. 63-71

Avraamidou, L., & Osborne, J. (2009). The Role of Narrative in Communicating Science. International Journal of Science Education, 31(12), 1683-1707

Balaton, M., Cavadas, J., Carvalho, P. S., & Lima, J. J. G. (2021). Programming Ozobots for teaching astronomy. Physics Education, 56(4), 045018

Bell, T., Alexander, J., Freeman, I., & Grimley, E. (2009). Computer science unplugged: School students doing real computing without computers, New Zealand Journal of Applied Computing and Information Technology, 13(1), 20-29

Bers, M. U., Flannery, L., Kazakoff, E. R., & Sullivan, A., (2014). Computational thinking and tinkering: Exploration of an early childhood robotics curriculum, Computers & Education, 72, 145-157

Bogliolo, A. (2017). Pixel art, coding e immagini digitali. https://codemooc.org/pixel-art/

Bogliolo, A. (2020) A scuola con CodyRoby. Giunti Scuola

Campbell, J. (1949). The Hero with a Thousand Faces. New World Library, 2008 Ed.

Capecchi, S., Gena, C. & Lombardi, I. (2022). Visual and unplugged coding with smart toys. AVI 2022: Proceedings of the 2022 International Conference on Advanced Visual Interfaces

Carroll, B. W. & Ostlie, D. A. (1996). An Introduction to Modern Astrophysics, Pearson Education.

Di Blas, N., Paolini, P., Sabiescu, A. (2010). Collective digital storytelling at school as a whole-class interaction. IDC '10: Proceedings of the 9th International Conference on Interaction Design and Children, 11–19

Eguchi A. (2010). What is educational robotics? Theories behind it and practical implementation, in Gibson D. & Dodge B., eds., Proceedings of Society for Information Technology & Teacher Education International Conference 2010 (pp. 4006-4014), AACE, Chesapeake, VA.

Grover, S. & Pea, R. D. (2013). Computational Thinking in K–12. A Review of the State of the Field, Educational Researcher, 42(1), 38-43

Gulay, H. (2010). An earthquake education program with parent participation for preschool children. Educational Research and Review Vol. 5 (10), 624-630

Jenkins, H. (2009). Confronting the Challenges of Participatory Culture: Media Education for the 21st Century. The MIT Press, Cambridge, MA

Kazakoff, E. R., Sullivan, A., & Bers, M. U. (2013). The Effect of a Classroom-Based Intensive Robotics and Programming Workshop on Sequencing Ability in Early Childhood, Early Childhood Education Journal, 41, 245-255

Lee, J., Junoh, J. (2019). Implementing unplugged coding activities in early childhood classrooms, Early Childhood Education Journal, 47, 709-716

Mangione, G. R., Orciuoli, F., Pierri, A., Ritrovato, P., Rosciano, M. (2011). A New Model for Storytelling Complex Learning Objects. 2011 Third International Conference on Intelligent Networking and Collaborative Systems, Fukuoka, Japan, 2011, pp. 836-841

Mangione, G. R., Capuano, N., Orciuoli, F., & Ritrovato, P. (2013). Disaster Education: a narrative-based approach to support learning, motivation and students' engagement. Journal of E-Learning and Knowledge Society, 9(2)

Maragliano, R., & Pireddu, M. (2015). Ripensare il medium didattico. Mediascapes Journal, (5), 3-11

Peduto, C. (2012). The fantastic voyage of Nino the neutrino, from the Sun to the Earth. CERN Report IPPOG-RDB-2022-050

Pekarova, J. (2008). Using a Programmable Toy at Preschool Age: Why and How? Workshop Proceedings of SIMPAR 2008 Intl. Conf. on Simulation, Modeling And Programming For Autonomous Robots, Venice (Italy)

Petrucco, C., & De Rossi, M. (2009). Narrare con il digital storytelling a scuola e nelle organizzazioni. Carocci Editore, Roma

Plass, J. L., Homer, B. D., & Kinzer, C. K. (2015). Foundations of game-based learning. Educational Psychologist, 50(4), 258–283

Rivoltella, P. C. (2012). Neurodidattica. Insegnare al cervello che apprende. Raffaello Cortina, Milano

Rusk, N., Resnick, M., Berg, R., & Pezalla-Granlund, M. (2008). New pathways into robotics: Strategies for broadening participation. Journal of Science Education and Technology, 17(1), 59–69

Sandri, M., Mancini, D., Ammoscato, A., Catania, M., D'Alessio, F., Bondesan, B., Di Giacomo, F., Fulco, M. T., Galleti, S., Leonardi, L., Leoni, R., Mignone, C., & Russo, R. (2023). Educational robotics as a powerful tool to create motivating activities in educational environments. Memorie della Società Astronomica Italiana, 94, 120

Souza, M. A. M., & Duarte, J. R. R. (2015). Low-cost educational robotics applied to physics teaching in Brazil. Physics Education, 50, 4

Vizzari, A. M. (2022). Didattica con le Escape Room. Spunti metodologici e percorsi operativi multidisciplinari. Scuola primaria e secondaria di primo grado. Erickson, Trento

Wing, J. M. (2006). Computational thinking. Communications of the ACM, 49(3):33–35

www.ingramcontent.com/pod-product-compliance
Lightning Source LLC
Chambersburg PA
CBHW071113050326
40690CB00008B/1211